T0205501

Welding Simulations Using ABAQUS

Bahman Meyghani · Mokhtar Awang

Welding Simulations Using ABAQUS

A Practical Guide for Engineers

 Springer

Bahman Meyghani
BKL B.V.
Collse Heide 1
Nuenen, The Netherlands

Department of Mechanical Engineering
Faculty of Engineering Technology
and Built Environment, UCSI University
Kuala Lumpur, Malaysia

Mokhtar Awang
Mechanical Engineering Department
Universiti Teknologi Petronas
Seri Iskandar, Perak, Malaysia

ISBN 978-981-19-1322-8   ISBN 978-981-19-1320-4 (eBook)
https://doi.org/10.1007/978-981-19-1320-4

This Springer imprint is published by the registered company Springer Nature Singapore Pte Ltd.
The registered company address is: 152 Beach Road, #21-01/04 Gateway East, Singapore 189721,
Singapore

To our beloved families and friends.
Thank you for all years of love,
encouragement and unfailing support.

Preface

Increasing human needs leads to the creation of complicated scientific problems. Solving a lot of mechanical and industrial problems requires the use of advanced joining methods to join different parts, components, similar and dissimilar metals or some nonmetal materials together. Understanding the behavior of these advanced joining methods by using classical computational methods such as elasticity theory of stress distribution and classical temperature distribution governing equations is challenging. To overcome the difficulties of these challenges, finite element methods are proposed as an appropriate method. In this regard, ABAQUS is well known as a user-friendly commercial finite element software for modeling different processes in mechanical, civil, aerospace and some other engineering fields. As an advanced welding method, Friction Stir Welding (FSW) has a complicated thermomechanical nature, and investigating the process behavior has attracted enormous research interests during the past decades. Numerical simulation is one of the most important methods for digging down deep into the process in detail. This tutorial book contains unified and detailed tutorials for engineers and students who are interested in simulating the FSW process in ABAQUS finite element software.

Nuenen, The Netherlands/Kuala
Lumpur, Malaysia
Seri Iskandar, Malaysia

Bahman Meyghani
Mokhtar Awang

Acknowledgements The authors of this book would like to acknowledge Prof. Wallace Kaufman for his endless support.

About ABAQUS Modules

In this section, it is important to explain ABAQUS software modules. The part module is normally used for creating the geometries or importing complicated geometries from other cad software. The property module is used to assign the material, create the section and assign the section to a part. Assembly module is used to instance the part meaning. Step module is used for creating the simulation steps. Interaction module is employed to create the interactions and contact behavior between different parts. Load module is used to define the loads, boundary conditions and predefined parameters such as initial temperature. Finite element meshes are defined in the mesh module. Job module is used to create the input file and define the solver. Visualization module is used to visualize the simulation results. Sketch module is almost the same as part module; however, it can be only employed for sketching.

Contents

List of Figures

Chapter 1
Introduction to Friction Stir Welding (FSW)

Friction stir welding (FSW) (Fig. 1.1) is a solid-state joining processes in which there is a complicated thermomechanical behavior [1]. Modelling and simulation of this processes has been defined as an interesting research topic for enormous research interests during the past decade, however researchers always face a lot of challenges [2–4]. Some of these challenges are nonlinear thermomechanical behavior, mesh distortion, complicated contact behavior, etc. [5].

ABAQUS software in known as a powerful Finite Element (FE) package for simulating FSW [6–8]. There are various methods for simulating the FSW process in ABAQUS. All of these methods have their own advantages and disadvantages [9]. Therefore, based on the required outcome researchers select the appropriate modelling method. However, studying all of these methods requires a lot computational and economical costs [10].

In this book, different FSW simulation methods are summarized in order to assist researchers to learn the process modelling fast and appropriate [11, 12]. In each section, all of the required steps are explained in detail [13, 14]. In the first part of the tutorial chapters the preprocessor section is described like the geometry, material property definition, etc. In the second part, the postprocessor like the visualization of the results, etc. are explained [15–17].

© The Author(s), under exclusive license to Springer Nature Singapore Pte Ltd. 2022
B. Meyghani and M. Awang, *Welding Simulations Using ABAQUS*,
https://doi.org/10.1007/978-981-19-1320-4_1

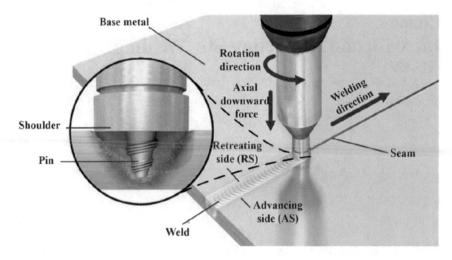

Fig. 1.1 Schematic view of the FSW process [18]

References

1. B. Meyghani, Thermomechanical analysis of friction stir welding (FSW) on curved plates by adapting calculated temperature dependent properties (Universiti Teknologi Petronas, 2018)
2. S. Emamian, M. Awang, P. Hussai, B. Meyghani, A. Zafar, Influences of tool pin profile on the friction stir welding of AA6061. ARPN J. Eng. Appl. Sci. **11**(20), 12258–12261 (2016)
3. B. Meyghani, M.B. Awang, M. Momeni, M. Rynkovskaya, Development of a finite element model for thermal analysis of friction stir welding (FSW), in *IOP Conference Series: Materials Science and Engineering*, vol. 495, no. 1 (IOP Publishing, 2019), p. 012101
4. B. Meyghani, M.B. Awang, R. Teimouri, Prediction of the temperature behaviour during friction stir welding (FSW) using hyperworks® (Springer Singapore, Singapore, 2021), pp. 119–130
5. B. Meyghani, M. Awang, S. Emamian, M.K.B.M. Nor, Thermal modelling of friction stir welding (FSW) using calculated Young's modulus values, in *The Advances in Joining Technology* (Springer, 2018), pp. 1–13
6. S.S. Emamian, M. Awang, F. Yusof, *Advances in Manufacturing Engineering: Selected Articles from ICMMPE 2019* (Springer Nature, 2020)
7. B. Meyghani, M. Awang, Developing a finite element model for thermal analysis of friction stir welding (FSW) using hyperworks, in *Advances in Material Sciences and Engineering* (Springer, 2020), pp. 619–628
8. B. Meyghani, M. Awang, S. Emamian, B. Plank, C. Heinzl, K. Siow, Stress analysis of nano porous material using computed tomography images. Materialwiss. Werkstofftech. **50**(3), 234–239 (2019)
9. B. Meyghani, M. Awang, S. Emamian, A mathematical formulation for calculating temperature dependent friction coefficient values: application in friction stir welding (FSW), in *Defect and Diffusion Forum,* vol. 379 (2017), pp. 73–82
10. B. Meyghani, M.B. Awang, R.G.M. Poshteh, M. Momeni, S. Kakooei, Z. Hamdi, The effect of friction coefficient in thermal analysis of friction stir welding (FSW), in *IOP Conference Series: Materials Science and Engineering*, vol. 495, no. 1 (IOP Publishing, 2019), p. 012102
11. B. Meyghani, M. Awang, C.S. Wu, S. Emamian, Temperature distribution investigation during friction stir welding (FSW) using smoothed-particle hydrodynamics (SPH), in *Advances in Manufacturing Engineering* (Springer, 2020), pp. 749–761

12. M. Awang, S.S. Emamian, F. Yusof, *Advances in Material Sciences and Engineering* (Springer, 2020)
13. S. Emamian, M. Awang, F. Yusof, P. Hussain, B. Meyghani, A. Zafar, The effect of pin profiles and process parameters on temperature and tensile strength in friction stir welding of AL6061 alloy, in *The Advances in Joining Technology* (Springer, 2019), pp. 15–37
14. B. Meyghani, M. Awang, S. Emamian, E. Akinlabi, A comparison between temperature dependent and constant young's modulus values in investigating the effect of the process parameters on thermal behaviour during friction stir welding: vergleich zwischen den temperaturabhängigen und konstanten elastizitätsmodulwerten in der untersuchung der prozessparameter auf die wärmewirkung beim rührreibschweißen. Materialwiss. Werkstofftech. **49**(4), 427–434 (2018)
15. W. Sutopo et al., *IOP Conference Series: Materials Science and Engineering* (2019)
16. M. Awang, *The Advances in Joining Technology* (Springer, 2019)
17. B. Meyghani, M.B. Awang, R. Teimouri, Prediction of the temperature behaviour during friction stir welding (FSW) using hyperworks®, in *Advances in Material Science and Engineering* (Springer, 2021), pp. 119–130
18. B. Meyghani, C. Wu, Progress in thermomechanical analysis of friction stir welding. Chin. J. Mech. Eng. **33**(1), 12 (2020)

Chapter 2
Thermomechanical Analysis

Thermomechanical analysis (TMA) is a modeling technique that studies the properties of the material with respect to the temperature. It is very significant because a lot of material properties change during the heating and cooling phases. Basically, FSW is a temperature dependent process in which the welding quality highly depends on the process temperature. Therefore, thermomechanical study of FSW assists in investigating the processes more in detail [1, 2]. To illustrate, TMA is a fundamental tool in material science, specifically in thermal behavior investigation. TMA states that the changes of temperature can have an influence on the material behavior [3]. In this regard, FSW TMA could facilitate a further investigation on the process and consequently help improve the welding quality for the similar or dissimilar materials. To clarify the influence of the temperature on the welding quality some researchers are done by the literature. For example, it is claimed that across the same material, the distribution of the heat is highly asymmetrical or it was investigated that the welding longitudinal residual stress rises as the tool movement and the welding speed increases (both of these speeds affect the welding temperature) [4].

To explain more, plastic deformation and frictional force are two main sources for the heat generation. Past studies [2, 5] claimed that the maximum temperature in this process could reach up to 60%-80% of the melting temperature of the base material. Thus, the FSW process should be modelled as a fully coupled thermomechanical analysis. Hence, for simulating the FSW process many studies [6, 7] had adapted temperature dependent material properties that were obtained from the experimental tests.

Fundamentally variations of the temperature will affect the behavior of the material like the elasticity, viscoelasticity and viscous. For instance, if the temperature control is poor [8] the material will be completely unusable. Thus, if the welding temperature becomes very high or very low, the material microstructure will be altered. According to the aforesaid descriptions, FSW is a temperature-dependent process in which the quality of the welding depends on the temperature.

Normally, during the experimental investigation the temperature is recorded by using thermocouples and the stress can be measured by the creep and the stress

© The Author(s), under exclusive license to Springer Nature Singapore Pte Ltd. 2022   5
B. Meyghani and M. Awang, *Welding Simulations Using ABAQUS*,
https://doi.org/10.1007/978-981-19-1320-4_2

relaxation. Control force tools can be also used for measuring the thermomechanical behavior. As a drawback, using the experimental tools for measuring the thermome-chanical behavior during FSW are limited and costs a lot of money. Therefore, exper-imental measurement of the yield and the break properties for different specimens is not accurately doable and costs a lot.

As mentioned, by using TMA, the behavior of FSW with the change of the temper-ature will be investigated. In this regard, FSW TMA could facilitate a further inves-tigation on the process and consequently assist in enhancing the welding quality for both similar or dissimilar materials [4].

References

1. B. Meyghani, M. Awang, S. Emamian, B. Plank, C. Heinzl, K. Siow, Stress analysis of nano porous material using computed tomography images. Materialwiss. Werkstofftech. **50**(3), 234–239 (2019)
2. B. Meyghani, M.B. Awang, Prediction of the temperature distribution during friction stir welding (Fsw) with a complex curved welding seam: application in the automotive industry. MATEC Web Conf. **225**, 01001 (2018)
3. M. Awang, *2nd International Conference on Mechanical, Manufacturing and Process Plant Engineering* (Springer, 2017)
4. S. Gaisford, V. Kett, P. Haines, *Principles of Thermal Analysis and Calorimetry* (Royal society of chemistry, 2016)
5. M. Awang, S.R. Khan, B. Ghazanfar, F.A. Latif, Design, fabrication and testing of fixture for implementation of a new approach to incorporate tool tilting in friction stir welding, in *MATEC Web of Conferences*, vol. 13 (EDP Sciences, 2014), p. 04020
6. D. Maisonnette, D. Bardel, V. Robin, D. Nelias, M. Suery, Mechanical behaviour at high temper-ature as induced during welding of a 6xxx series aluminium alloy. Int. J. Press. Vessels Pip. **149**, 55–65 (2017)
7. B. Meyghani, M. Awang, S. Emamian, A comparative study of finite element analysis for friction stir welding application. ARPN J. Eng. Appl. Sci. **11**(22), 12984–12989 (2016)
8. L. Cederqvist, O. Garpinger, T. Hagglund, A. Robertsson, Reliable sealing of copper canisters through cascaded control of power input and tool temperature, in *Friction Stir Welding and Processing VI* (2011), pp. 49–58

Chapter 3
Finite Element Modeling of FSW

Finite element methods (FEMs) is one of the most important approaches for analyzing the structures. In these methods, the discrete of the matrix equation systems will be done for finding the influence of the mass and the stiffness in the continuous structure. In these methods, the symmetric matrices need to be banded. The problem of the complex geometry of the domain will be approximately solved, because the mass and the stiffness matrices will be assembled from a finite element model in which the individual domain will be divided into simple shapes. Therefore, in a simple geometrical description a mathematical formulation will be used regardless of the overall structure geometry. To illustrate more, in FEMs the structure will be divided to discrete regions or volumes called elements. The boundaries of the elements are connected by some points called nodes. In the most common type of the element a same polynomial will be used for the displacements of the elements to the node displacement. This processes named as the shape function interpolation.

FEMs have a wide range of the application in simulating different processes such as thermal behavior, fluid behavior, wave propagation, the deformation behavior, etc. Most of the abovementioned analysis use the partial differential equations (PDFs) which can be solved by using numerical techniques like FEMs. Therefore, the application of FEMs is increasing in many industries like aerospace, shipping, shipbuilding, marine etc. Furthermore, coupling different problems together like thermomechanical, fluid structures, thermo-chemo-mechanical thermochemical etc. are now widely using to solve complicated problems. Numerical models can be complemented by inverse analysis and this analysis can be validated and identified by using experimental parameters. Previous literature used these techniques to solve complicated frictional behaviors and the parameters for inertial friction and friction stir welding simulations. In this regard, inverse analysis can be used to reduce the difference between the experimental and the simulation results [1, 2].

Therefore, numerical modelling of FSW is a very complicated task, therefore it is studied by many researchers [1, 2]. However, the complexity of the process modelling can be solved by using finite element models. Since FSW involves changes in the temperature and the mechanical behavior of the material, couple thermomechanical

© The Author(s), under exclusive license to Springer Nature Singapore Pte Ltd. 2022
B. Meyghani and M. Awang, *Welding Simulations Using ABAQUS*,
https://doi.org/10.1007/978-981-19-1320-4_3

modelling can be used to investigate the FSW deeply. To illustrate, the temperature and the material flow etc. can be simulated by using numerical methods, thus the welding behavior can be investigated more in detail and this issue assists in the enhancement of the welding quality. In addition, some software has the capability to be modified. To explain the issue, user defined subroutines can be used in the software to define specific simulation behaviors like special material properties or a special contact condition. Consequently, FEMs can be used as an appropriate method for thermomechanical modelling of the FSW for similar and dissimilar materials.

3.1 Different Modelling Methods

Computational solid mechanics (CSM) and computational fluid dynamics (CFD) are two common modelling methods for FSW process. In CSM method, the workpiece is modelled as a body which is allowed to be deformed. In contrast, the tool is normally considered as a rigid body in which the deformation is not allowed. In this modelling, the friction at the tool and the workpiece interface is the main source of the heat during the welding.

In CFD method, the workpiece is normally considered as a non-Newtonian, incompressible and visco-plastic material. CFD methods can be employed to explain the physical quantities. In this situation, the mesh grid is fixed in the domain and the grid distortion is avoided. Thus, it has some advantages in dealing with the complex geometries.

3.1.1 CSM Method

The issue of interest in CSM based methods for FSW is investigating technical approaches for enhancing the modelling of the process. As an example, achieving better qualities for the mesh is important because the processes undergoes large plastic deformation. Accordingly, a three-dimensional mesh that can be fined for complicated geometries is needed for the thermomechanical examination of the FSW process. In addition, the mesh quality is exceptionally affected by the forecast thermomechanical state variable values and the distributions, because the mesh quality has a significant effect on the convergence of the simulation and the computational robustness. In a general thermomechanical simulation which is performed based on CSM, the development of the mesh is related to the material behavior and the material deformation can be investigated by the mesh deformation. During FSW, the extreme mesh distortion would cause the crashing of the calculation. Thus, to have a high accuracy thermomechanical modelling of FSW based on CSM, it is essential to pick the right way to deal with deal with the mesh distortion framed by the serious plastic deformity [3].

The research in this area demonstrates a few favorable methods for controlling distortion of the mesh. One of the solutions for this issue is using arbitrary Lagrangian Eulerian (ALE) method [4–7]. Fixing or attaching the mesh based on the deformations is doable in the ALE technique, thus the material is able to move across the mesh. Another proposed method is using Lagrangian domain with intermittent remeshing [8–10]. To explain more, in the Lagrangian method the material is able to move across the mesh and the re-meshing can be done regularly. It should be mentioned that, the re-meshing can perform a high quality evaluation of the thermomechanical behavior during the computation modelling. Furthermore, FSW has a dynamic nature, therefore the workpiece material needs to be considered as a rate-dependent properties. As an alternative, the CSM explicit central difference method can be normally used for integrating the equations in time. Another good alternative method for thermomechanical behavior is the coupled Eulerian–Lagrangian (CEL) or pure Eulerian methods [11]. Many studies investigated the FSW and some other analogous processes by using the CEL method [12–14]. The basis of the Eulerian formulation is discretized in the CEL method for the workpiece, and a Lagrangian domain is considered as the welding tool. It needs to be mentioned, on the basis of the immersed boundary method, the implementation of the tool/workpiece interaction is done in the CEL approach [15]. In the pure Eulerian method, the workpiece mesh remains constant and the material can be easily move across the mesh [16]. Therefore, it can be considered as an alternative method for investigating the thermomechanical behavior and also the material flow during the process.

3.1.2 CFD Method

Based on CFD method, some thermomechanical analysis of FSW [17–19] is done. In these models, the analysis of the 3-D heat transfer and flow of the material were done by resolving mass conservation, momentum, and energy equations [20]. In an Eulerian based CFD method, during the material flow or the movement of the material, the computational mesh will not deform. Thus, the movement of the material is not depended to the mesh. In this method, the initial discretization steps of the analysis with a constant mesh is employed. After that, a dynamic moving mesh method is used. In the abovementioned method of the simulation, a dynamic geometric model allows the acquire transient behavior. A CFD-based analysis was proposed by Yu et al. [21] for examining the attributes of the heat transient flow and the mass equations. In the research, the material could move across the mesh including the geometric model variation, the transitional and the tool rotating movements. A reference [22] permitted the mesh revolution in examining the impact of the pin profile during the thermomechanical analysis of FSW. A CFD-based analysis was also recommended for inspecting the generation of the heat, temperature distribution, and material movement [23]. Another literature [24] used a moving mesh technique for finding the pin thread impact on the flow of the material. In the paper, the material which is located in the pin thread channels is considered as an extra fluid volume moving mesh. Moreover, in the CFD-based FSW analysis, a transient phenomenon is considered for the moving mesh approaches.

3.1.3   Comparison of Governing Equations for CSM and CFD Methods

As mentioned earlier, a category of computational mechanics that use the computational methods for examining the situation under the control of the solid mechanics materials called computational solid mechanical (CSM). This method is able to model high-temperature severe plastic deformation, model forming, and thermo-mechanical condition in material manufacturing processes [25, 26]. Elasto-viscoplastic theory dominates the simulation during the CSM modelling. To illustrate, by using CSM the governing equations can be solved to obtain thermal and mechanical responses. The differential form of the motion-equation can be employed to solve the displacement vector which is one the governing equation in the CSM. Conversely, a non-Newtonian fluid should be considered for the workpiece in CFD models. The analysis of the thermo-mechanical attributes in FSW employs the conservation equations of mass, momentum, and energy. It should be noted that, in the CFD-based analysis, the simple form of the governing equations for the material movement without considering the elastic term is used as the momentum conservation equation.

3.1.4   Computational Approaches

It is anticipated to find and calculate thermomechanical variables distributions with a suitable spatial determination. In the computational examination, spatial determination relies on the mesh size. It can be summarized that in the CSM-based analyses, the region near the tool has more attention, thus the mesh size near the tool area was found to be in the range of 0.2–1.0 mm [9, 27–30]. For CFD based analyses the values of 0.1–0.2 mm were reported [31, 32]. Based on the recent publications, it can be seen that there is a significant change for the microstructure almost 0.1 mm away from the stirring zone [3, 33]. Therefore, the mesh size should be equal or less than 0.1 mm. Even though a lot of methods were proposed for FSW analysis, however the bulk mesh ending up in deficient spatial resolution remains as an existing issue. Moving forward, more analysis on the basis of CSM model is needed to use smaller sizes for the mesh which leads to the enhancement of the accuracy. Thus, the level of the computation needs to be enhanced to permit an improved spatial resolution for developing thermomechanical analysis of FSW. At the same time, the computation of the equivalent strategies is needed for a tremendous increase in the computation efficiency and shorten the computational costs. A note is made that the presently accessible computational capacity permits the employment of smaller size for the mesh during CFD-based simulations (in comparison with the CSM based modelling). This issue happens in the CFD-based models due to the simplification of the governing equation which has neglected the terms of the elasticity. Furthermore, the stress calculations are only dependent on the velocity gradient, therefore in a large extent, the computational costs for the CFD models are less than CSM models [34]. Presented CFD-based FSW analysis permits the improvement of the spatial resolution, hence much complicated geometrical features can be simulated by CFD based models [24]. Whereas, the mean stress also called the forecast pressure has more ambiguity in the CFD based analyses, due to the negligence of elasticity [3].

In the initial steps of the welding simulation (plunging step), the substantially stable time increments is less than the average values. This issue happening because the pin elements are less than the shoulder. These issues results in small-time increment for the entire structure. It is also clear that, increasing the number of the elements rises the calculation time. Normally, solving the governing equations by the explicit algorithm takes more time. Hence, for decreasing simulation time, the element mass needs to be improved. In this case, the 'scaling procedure' is one of the most appropriate solutions for decreasing the simulation time, because the mass of the matrix plays a significant role in computational efficiency and the model accuracy.

References

1. L. D'Alvise, E. Massoni, S. Walløe, Finite element modelling of the inertia friction welding process between dissimilar materials. J. Mater. Process. Technol. **125**, 387–391 (2002)
2. C. Bennett, M. Attallah, M. Preuss, P. Shipway, T. Hyde, S. Bray, Finite element modeling of the inertia friction welding of dissimilar high-strength steels. Metall. Mater. Trans. A **44**(11), 5054–5064 (2013)
3. G. Chen, S. Zhang, Y. Zhu, C. Yang, Q. Shi, Thermo-mechanical analysis of friction stir welding: a review on recent advances. Acta Metall. Sinica (English Letters) 1–10
4. H. Schmidt, J. Hattel, A local model for the thermomechanical conditions in friction stir welding. Modell. Simul. Mater. Sci. Eng. **13**(1), 77 (2004)
5. Z. Zhang, Q. Wu, Numerical studies of tool diameter on strain rates, temperature rises and grain sizes in friction stir welding. J. Mech. Sci. Technol. **29**(10), 4121–4128 (2015)
6. Z. Zhang, Y. Liu, J. Chen, Effect of shoulder size on the temperature rise and the material deformation in friction stir welding. Int. J. Adv. Manuf. Technol. **45**(9–10), 889 (2009)
7. Z. Zhang, J. Chen, The simulation of material behaviors in friction stir welding process by using rate-dependent constitutive model. J. Mater. Sci. **43**(1), 222–232 (2008)
8. G. Buffa, J. Hua, R. Shivpuri, L. Fratini, A continuum based fem model for friction stir welding—model development. Mater. Sci. Eng. A **419**(1–2), 389–396 (2006)
9. Z. Wan, Z. Zhang, X. Zhou, Finite element modeling of grain growth by point tracking method in friction stir welding of AA6082-T6. Int. J. Adv. Manuf. Technol. **90**(9–12), 3567–3574 (2017)
10. L. Long, G. Chen, S. Zhang, T. Liu, Q. Shi, Finite-element analysis of the tool tilt angle effect on the formation of friction stir welds. J. Manuf. Process. **30**, 562–569 (2017)
11. F. Al-Badour, N. Merah, A. Shuaib, A. Bazoune, Coupled Eulerian Lagrangian finite element modeling of friction stir welding processes. J. Mater. Process. Technol. **213**(8), 1433–1439 (2013)

12. K. Li, F. Jarrar, J. Sheikh-Ahmad, F. Ozturk, Using coupled Eulerian Lagrangian formulation for accurate modeling of the friction stir welding process. Procedia Eng. **207**, 574–579 (2017)
13. V. Shokri, A. Sadeghi, M. Sadeghi, Thermomechanical modeling of friction stir welding in a Cu-DSS dissimilar joint. J. Manuf. Process. **31**, 46–55 (2018)
14. X. Liu, S. Lan, J. Ni, Thermal mechanical modeling of the plunge stage during friction-stir welding of dissimilar Al 6061 to TRIP 780 steel. J. Manuf. Sci. Eng. **137**(5), 051017 (2015)
15. L. Zhu, C.S. Peskin, Simulation of a flapping flexible filament in a flowing soap film by the immersed boundary method. J. Comput. Phys. **179**(2), 452–468 (2002)
16. B. Meyghani, M.B. Awang, S.S. Emamian, M.K.B. Mohd Nor, S.R. Pedapati, A comparison of different finite element methods in the thermal analysis of friction stir welding (FSW). Metals **7**(10), 450 (2017)
17. P.A. Colegrove, H.R. Shercliff, 3-Dimensional CFD modelling of flow round a threaded friction stir welding tool profile. J. Mater. Process. Technol. **169**(2), 320–327 (2005)
18. P. Colegrove, H. Shercliff, CFD modelling of friction stir welding of thick plate 7449 aluminium alloy. Sci. Technol. Weld. Joining **11**(4), 429–441 (2006)
19. R. Nandan, G. Roy, T. Lienert, T. DebRoy, Numerical modelling of 3D plastic flow and heat transfer during friction stir welding of stainless steel. Sci. Technol. Weld. Joining **11**(5), 526–537 (2006)
20. B. Meyghani, A modified friction model and its application in finite-element analysis of friction stir welding process. J. Manuf. Process. **72**, 29–47, 2021/12/01 (2021)
21. Z. Yu, W. Zhang, H. Choo, Z. Feng, Transient heat and material flow modeling of friction stir processing of magnesium alloy using threaded tool. Metall. Mater. Trans. A **43**(2), 724–737 (2012)
22. H. Su, C.S. Wu, M. Bachmann, M. Rethmeier, Numerical modeling for the effect of pin profiles on thermal and material flow characteristics in friction stir welding. Mater. Des. **77**, 114–125 (2015)
23. L. Shi, C. Wu, Transient model of heat transfer and material flow at different stages of friction stir welding process. J. Manuf. Process. **25**, 323–339 (2017)
24. G. Chen et al., Effects of pin thread on the in-process material flow behavior during friction stir welding: a computational fluid dynamics study. Int. J. Mach. Tools Manuf **124**, 12–21 (2018)
25. A.E. Tekkaya, State-of-the-art of simulation of sheet metal forming. J. Mater. Process. Technol. **103**(1), 14–22 (2000)
26. J. Zhou, L.-X. Li, J. Duszczyk, Computer simulated and experimentally verified isothermal extrusion of 7075 aluminium through continuous ram speed variation. J. Mater. Process. Technol. **146**(2), 203–212 (2004)
27. X. Yang et al., Numerical modelling and experimental investigation of thermal and material flow in probeless friction stir spot welding process of Al 2198–T8. Sci. Technol. Weld. Joining **23**(8), 704–714 (2018)
28. P. Chauhan, R. Jain, S.K. Pal, S.B. Singh, Modeling of defects in friction stir welding using coupled Eulerian and Lagrangian method. J. Manuf. Process. **34**, 158–166 (2018)
29. Z. Zhang, Q. Wu, M. Grujicic, Z. Wan, Monte Carlo simulation of grain growth and welding zones in friction stir welding of AA6082-T6. J. Mater. Sci. **51**(4), 1882–1895 (2016)
30. Z. Sun, C. Wu, A numerical model of pin thread effect on material flow and heat generation in shear layer during friction stir welding. J. Manuf. Process. **36**, 10–21 (2018)
31. G. Chen, Q. Ma, S. Zhang, J. Wu, G. Zhang, Q. Shi, Computational fluid dynamics simulation of friction stir welding: a comparative study on different frictional boundary conditions. J. Mater. Sci. Technol. **34**(1), 128–134 (2018)
32. H. Su, C.S. Wu, A. Pittner, M. Rethmeier, Thermal energy generation and distribution in friction stir welding of aluminum alloys. Energy **77**, 720–731 (2014)
33. Y. Zhong, C. Wu, G. Padhy, Effect of ultrasonic vibration on welding load, temperature and material flow in friction stir welding. J. Mater. Process. Technol. **239**, 273–283 (2017)
34. P. Bussetta, É. Feulvarch, A. Tongne, R. Boman, J.-M. Bergheau, J.-P. Ponthot, Two 3D thermomechanical numerical models of friction stir welding processes with a trigonal pin. Numer. Heat Transf. Part A Appl. **70**(9), 995–1008 (2016)

Chapter 4
Arbitrary Lagrangian–Eulerian (ALE) Method

Here, the examination of the FSW simulation by using the ALE method is presented. This method is one of the most important methods for modelling FSW. To describe more, the Lagrangian and the Eulerian methods are not ideal for FSW local analysis (inside the welding zone) due to the occurrence of the divergence [1]. This is cased because the ALE method can solve the large plastic deformation by continuously remeshing at the local-level simulation while the Eulerian and Lagrangian method cannot (Fig. 4.1) [2, 3].

To explain this, in this technique the mesh on the region and the interfaces are free to move with the materials. As a general rule, ALE has a Lagrangian part for addressing deformations and an Eulerian part for tackling the advection issues of the velocity. Consequently, the significant target of this technique is solving the welding behavior inside the HAZ region [4]. At the local level simulation, the impact of the speed of the welding (transverse and rotational), the mechanisms of the contact, the shape of the pin, the size of the stir zone, and the material flow inside the stirring zone can be studied with a high accuracy [5, 6]. Moreover, the ALE strategy has an additional advantages over the Lagrangian and Eulerian techniques. It has an additional degree of freedom called the traditional kinematic depictions. This extra degree of freedom reduces the classical kinematic descriptions difficulties, because it joins the merits in an advantageous manner. These explanations make ALE proper for modelling the interface connections and furthermore for plastic deformations.

4.1 Geometry and Preprocessor

All basic definitions of the model (shapes, sections, etc.) make in the part module.

In order to start the modelling follow the following steps,

Part → Create Part.

Then, a name should be assigned for a 3-D, deformable, solid extrusion part (Fig. 4.2).

© The Author(s), under exclusive license to Springer Nature Singapore Pte Ltd. 2022
B. Meyghani and M. Awang, *Welding Simulations Using ABAQUS*,
https://doi.org/10.1007/978-981-19-1320-4_4

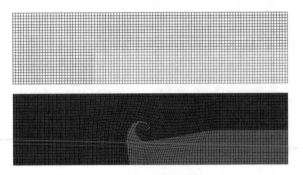

Fig. 4.1 Comparison between the main domain and the ALE

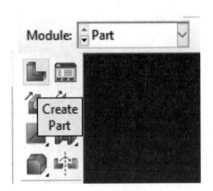

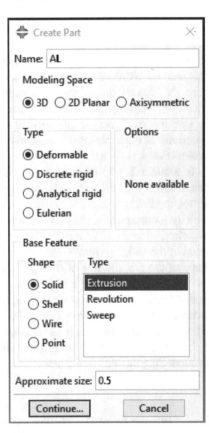

(a)   (b)

Fig. 4.2 Creating a new part

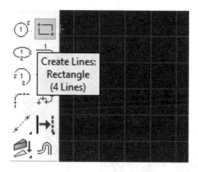

Fig. 4.3   Creating a rectangle

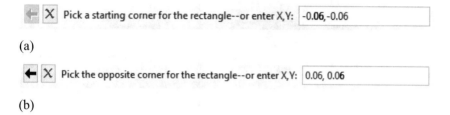

(a)

(b)

Fig. 4.4   Defining rectangle points

Create → Rectangle (Fig. 4.3).

Picking the starting point and the opposite corners (Fig. 4.4) and write down the following numbers.

After that, in order apply the changes the middle mouse button (scroll wheel) should be clicked two times (another way is to click on the Done order (the bottom of the page)).

Next, the thickness of 6 mm should be defined for the plate (Fig. 4.5).

After that the welding tool should be created. In the next step, the welding tool as a 3-D, discrete rigid, solid-revolutions part should be defined.

Fig. 4.5   Selecting material thickness

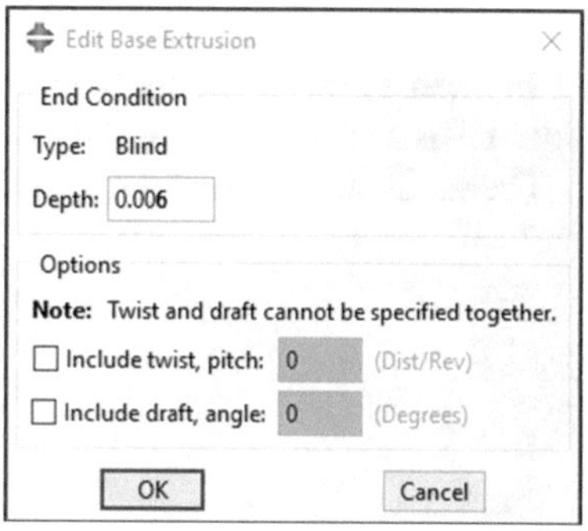

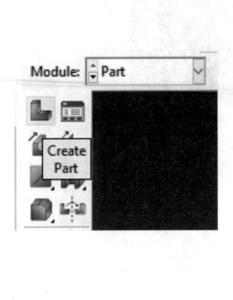
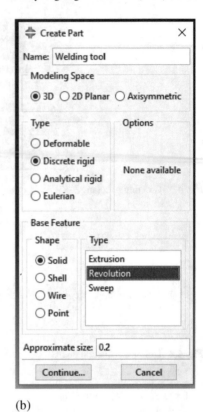

(a)   (b)

Fig. 4.6 Creating part

Part → Create (Fig. 4.6).

The dimensions for the tool are explained in Fig. 4.7.

Then, click on done at the below of page or single click on the middle mouse button (scroll wheel) and put 360 as the revolution angle (Fig. 4.8).

After that there is a need to define a reference point for the tool from the tools menu. This point can be placed anywhere in the tool (Fig. 4.9).

Tools → Reference point.

In order to avoid any problem in the assembly module, the tool should be also selected as a shell body (Figs. 4.10, 4.11 and 4.12).

Create Shell → Selecting shell from solid → Selecting the body → Click on done.

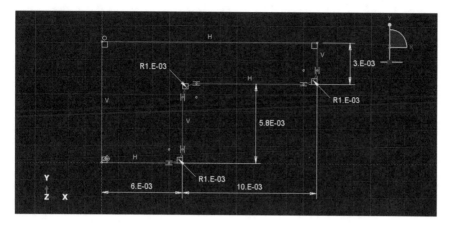

Fig. 4.7 Tool dimensions

Fig. 4.8 Revolving the tool

4.2 Material Property

In this module, the material property should be applied to the module (Fig. 4.13). The literature [7–13] values are also used for the material property and the interaction definitions.

For defining the material for the workpiece which is taken from the literature [14, 15] follow the below steps,

Material → Create.

- Thermal → Conductivity (Fig. 4.14).

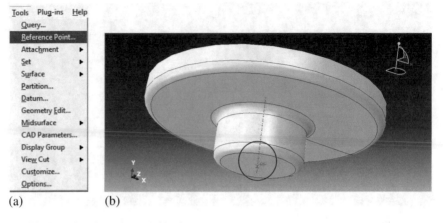

(a)   (b)

Fig. 4.9 Selecting the tool as a rigid reference

Fig. 4.10 Creating shell

Fig. 4.11 Selecting the tool

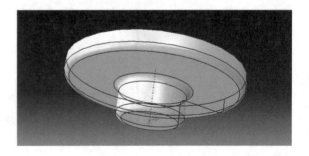

Fig. 4.12 Approving the
selected body

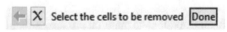

- General → Density (Fig. 4.15).

- Mechanical → Elasticity → Elastic (Fig. 4.16).

- Mechanical → Expansion (Fig. 4.17).

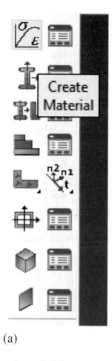

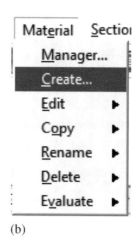

(a)   (b)

Fig. 4.13 Creating the material

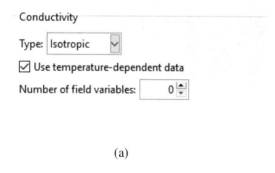

	Conductivity	Temp
1	135	25
2	177	100
3	184	150
4	192	200
5	200	250
6	210	300
7	220	400
8	230	450
9	265	500
10	285	550

(a)   (b)

Fig. 4.14 Defining thermal conductivity

- Thermal → Inelastic heat fraction (Fig. 4.18).

- Thermal → Latent heat (Fig. 4.19).

- Mechanical → Plasticity → Plastic (Fig. 4.20) [16].

  Rate dependent (Fig. 4.21).

- Thermal → Specific heat (Fig. 4.22).

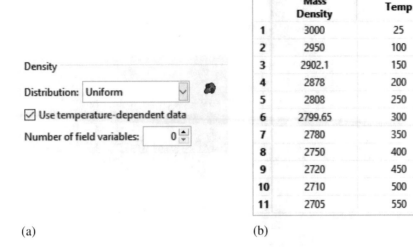

	Mass Density	Temp
1	3000	25
2	2950	100
3	2902.1	150
4	2878	200
5	2808	250
6	2799.65	300
7	2780	350
8	2750	400
9	2720	450
10	2710	500
11	2705	550

(a)   (b)

Fig. 4.15 Applying the density

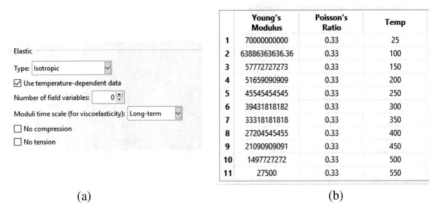

	Young's Modulus	Poisson's Ratio	Temp
1	70000000000	0.33	25
2	63886363636.36	0.33	100
3	57772727273	0.33	150
4	51659090909	0.33	200
5	45545454545	0.33	250
6	39431818182	0.33	300
7	33318181818	0.33	350
8	27204545455	0.33	400
9	21090909091	0.33	450
10	1497727272	0.33	500
11	27500	0.33	550

(a)   (b)

Fig. 4.16 Defining the elastic behaviour

The next step is creating a solid homogenous section.

Section → Create (Fig. 4.23).

The next step is assigning this section.

Assign → Section → Selecting the workpiece → Done → OK (Figs. 4.24, 4.25 and 4.26).

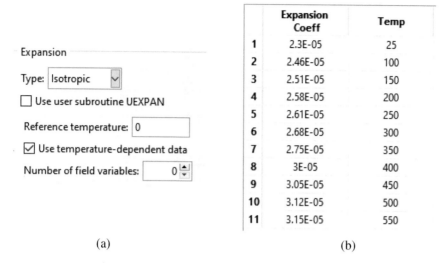

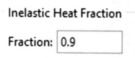

	Expansion Coeff	Temp
1	2.3E-05	25
2	2.46E-05	100
3	2.51E-05	150
4	2.58E-05	200
5	2.61E-05	250
6	2.68E-05	300
7	2.75E-05	350
8	3E-05	400
9	3.05E-05	450
10	3.12E-05	500
11	3.15E-05	550

(a)   (b)

Fig. 4.17 Defining the expansion

Inelastic Heat Fraction

Fraction: 0.9

Fig. 4.18 Selecting the inelastic heat fraction

Data

	Latent Heat	Solidus Temp	Liquidus Temp
1	2500	25	500

Fig. 4.19 Defining the latent heat

Plastic

Hardening: Johnson-Cook

Data

	A	B	n	m	Melting Temp	Transition Temp
1	369000000	684800000	0.73	1.72	500	25

Fig. 4.20 Defining the plasticity

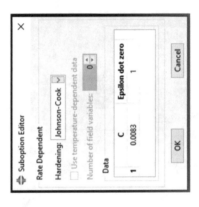

(b)

(a)

Fig. 4.21 Defining the rate dependent

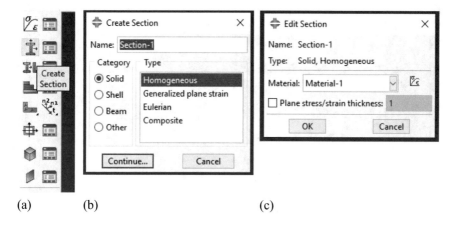

	Specific Heat	Temp
1	859.5	25
2	900.3	100
3	923.5	150
4	944.3	200
5	964.3	250
6	983.8	300
7	1003.5	350
8	1024	400
9	1046.1	450
10	1070.2	500
11	1097.2	550

(a)   (b)

Fig. 4.22 Defining the specific heat

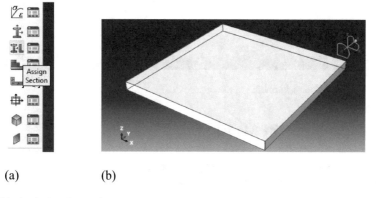

(a)   (b)   (c)

Fig. 4.23 Creating the solid homogenous section

(a)   (b)

Fig. 4.24 Assigning the section

Fig. 4.25 Selecting the region

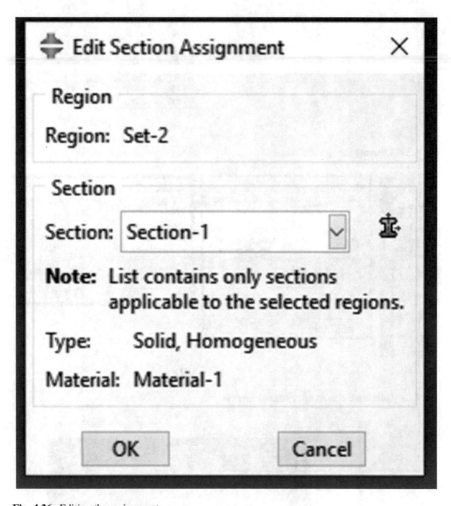

Fig. 4.26 Editing the assignment

4.3 Assembly of the Model

Instance → Create → Selecting bodies → OK (Fig. 4.27).

By following the below descriptions, the workpiece and the tool can be placed at a correct position (Figs. 4.28, 4.29 and 4.30).

Then, the tool should be moved upper by the following steps,

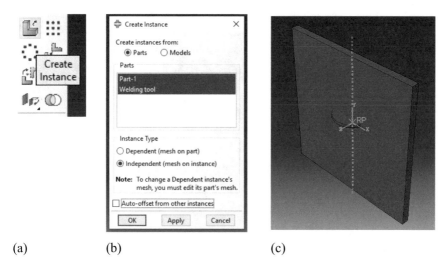

(a)   (b)   (c)

Fig. 4.27 Creating the assembly

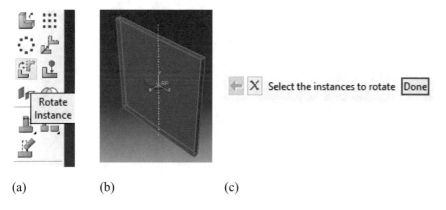

(a)   (b)   (c)

Fig. 4.28 Rotating the position

Translate instance → Selecting the tool → Done (Figs. 4.31, 4.32 and 4.33).

After that, for achieving a better and a high quality mesh the workpiece should be partitioned.

Tools → Partition → Face → Sketch selecting the face (Fig. 4.34).

Selecting the vertical line → Create lines rectangle (Fig. 4.35).

Draw a vertical rectangular to separate the welding zone from other parts (Fig. 4.36).

Then, double click on the middle mouse button (scroll wheel) to see Fig. 4.37.

To complete the partition click on tools → Partition → Cell → Extrude/ sweep edges → click on the middle mouse button (scroll wheel) → Select the rectangular

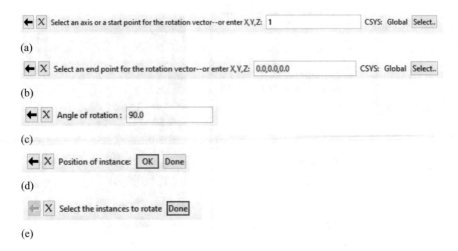

Fig. 4.29 Setting the position

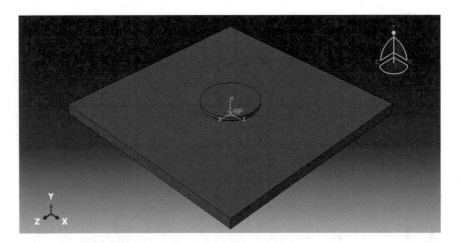

Fig. 4.30 Corrected position of the workpiece

edges → Done → Extrude along direction → Select the horizontal line → OK → Create partition (Figs. 4.38, 4.39, 4.40 and 4.41).

In the next part we have to do the same for another direction.

Tools → Partition → Face → Sketch selecting the face (Fig. 4.42).

Selecting the vertical line (Fig. 4.43).

Drawing a vertical rectangular to separate the welding zone from other parts (Fig. 4.44).

Then, double click on the middle mouse button (scroll wheel) to see Fig. 4.45.

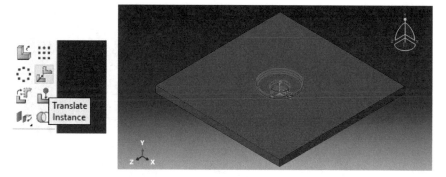

Fig. 4.31 Changing the position of the tool

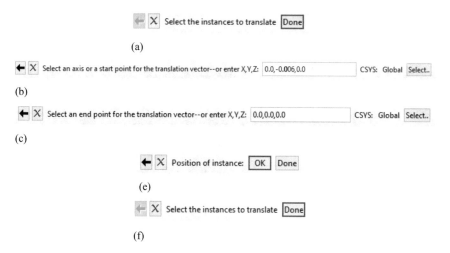

Fig. 4.32 Setting the tool position

To complete the partition click on tools → Partition → Cell → Extrude/ sweep edges → Click on the middle mouse button (scroll wheel) → Select three partitions → Click on the middle mouse button (scroll wheel) → Done (Figs. 4.46 and 4.47).

Then, Select the rectangular lines → Done → Extrude along direction → Select the horizontal line → OK → Create partition → Done (Figs. 4.48 and 4.49) (Figs. 4.50, 4.51, 4.52).

4.4 Step Creation

Step → Create (Figs. 4.53, 4.54 and 4.55).

The next step is to define Arbitrary Eulerian Lagrangian (ALE) modification.

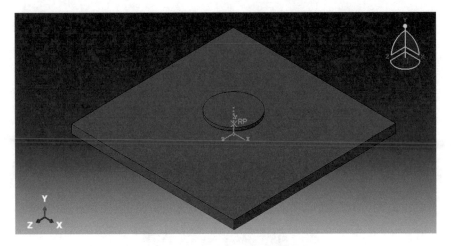

Fig. 4.33 Correction position for the tool and the workpiece

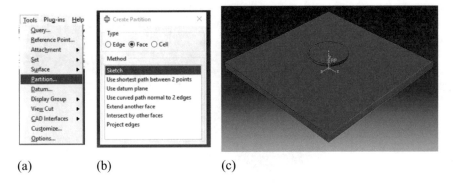

(a)   (b)   (c)

Fig. 4.34 Partitioning the workpiece

Fig. 4.35 Selecting the line
and creating the rectangle

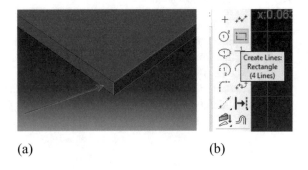

(a)   (b)

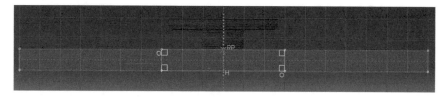

Fig. 4.36 Drawing the rectangle for the partitioning

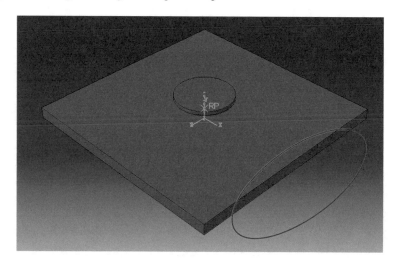

Fig. 4.37 Create partitions

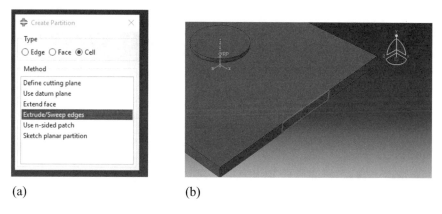

(a)   (b)

Fig. 4.38 Selecting the edge

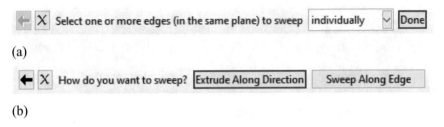

(a)

(b)

Fig. 4.39 Selecting the edges and extrude it along the direction

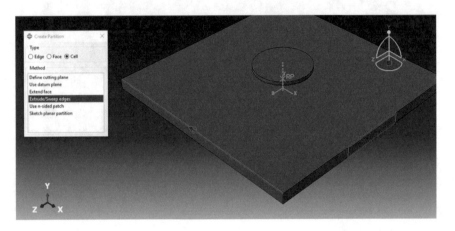

Fig. 4.40 Selecting the line

Fig. 4.41 Creating the partition

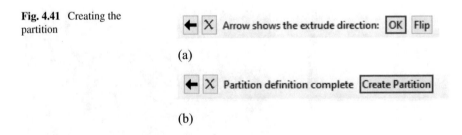

(a)

(b)

Other → ALE adaptive mesh control → Create → Continue → OK (Fig. 4.56).
Other → ALE adaptive mesh domain → Edit → Step-1 → Select the region → Select the center region → Done → OK (Figs. 4.57, 4.58 and 4.59).

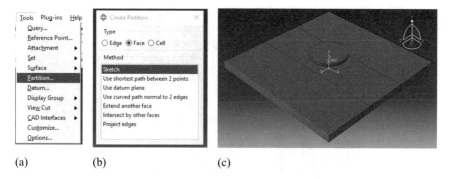

Fig. 4.42 Partitining the face

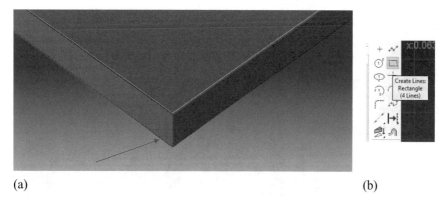

(a)   (b)

Fig. 4.43 Selecting the vertical line

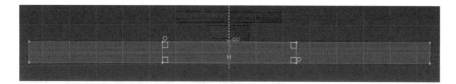

Fig. 4.44 Selecting the rectangle

4.5 Interactions

The first step is to create constraints in order to make the tool as a rigid body:

Constraint → Create Continue → Select the body → Edit selection (Fig. 4.60).

Select the tool → Select the predefined tool reference point → OK (Figs. 4.61, 4.62, 4.63 and 4.64).

.

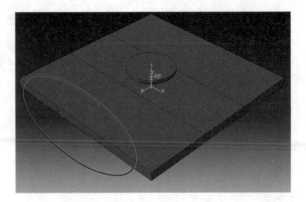

Fig. 4.45 Rectangle in another position

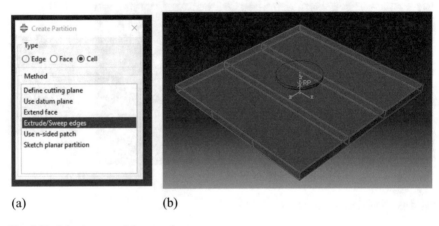

(a)   (b)

Fig. 4.46 Selecting extrude/sweep edges

Fig. 4.47 Selecting the cells

Fig. 4.48 Selecting the
rectangle

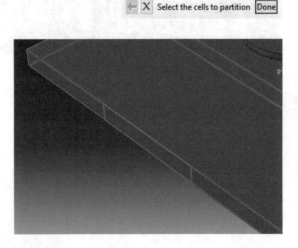

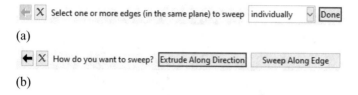

(a)

(b)

Fig. 4.49 Approving the selected region

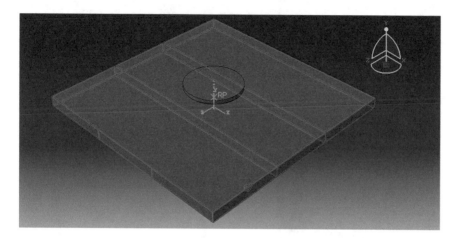

Fig. 4.50 Showing the preview of the partitioned region

Fig. 4.51 Creating the partition

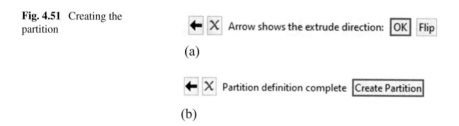

(a)

(b)

After that the interaction property should be created: Interaction → Property → Create → Select contact → Continue → Mechanical → Tangential behavior → Penalty → 0.3 (Fig. 4.65).

After that a hard contact should be defined:

Mechanical → Hard contact.

Then for the heat generation a contact should be define:

Thermal → Heat generation → Use the specify numbers → OK (Fig. 4.66).

The next step is to create interaction:

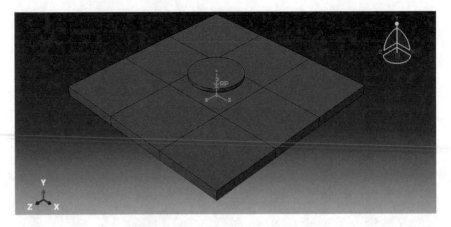

Fig. 4.52 Partitioned workpiece

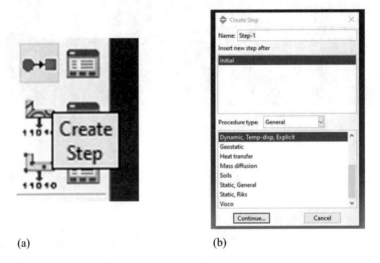

(a)   (b)

Fig. 4.53 Creating the step

Create → Interaction → Surface to surface contact → Continue → Select the bottom surfaces of the tool → Done → Brown → Surface → Done → Select the workpiece surface → Done → OK (Figs. 4.67, 4.68, 4.69, 4.70 and 4.71).

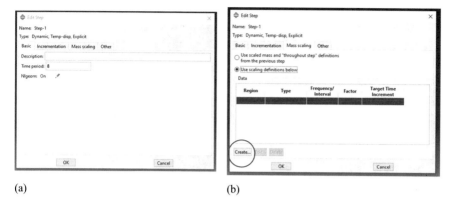

(a)   (b)

Fig. 4.54 Select the step time

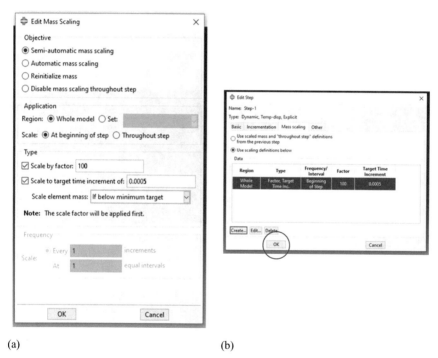

(a)   (b)

Fig. 4.55 Applying the mass scaling

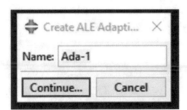

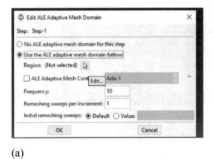

(a)   (b)

Fig. 4.56 Applying the ALE adaptive mesh

(a)   (b)

Fig. 4.57 Selecting ALE region

Fig. 4.58 Approving the region

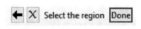

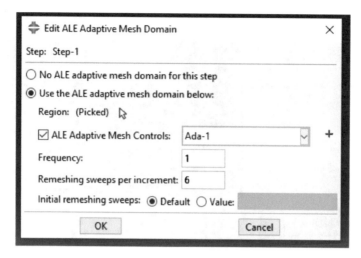

Fig. 4.59 Confirming the region

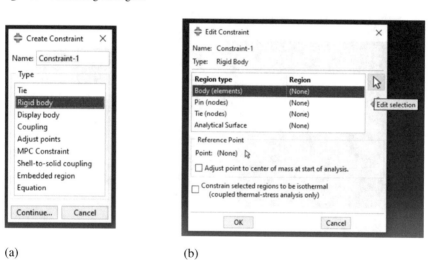

(a)   (b)

Fig. 4.60 Approving the rigid body

Fig. 4.61 Selecting the tool

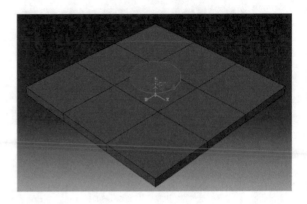

← X   Select body regions for the rigid body   (☑ Create set: b_Set-2   )   Done

Fig. 4.62 Confirming the region

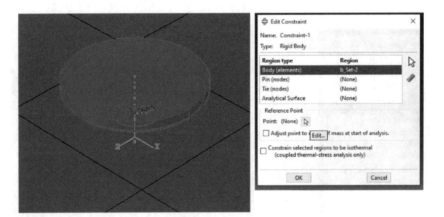

Fig. 4.63 Selecting the tool set

Fig. 4.64 Confirming the
rigid body

Fig. 4.65 Applying the
friction coefficient

Fig. 4.66 Applying the heat
generation

Edit Contact Property   ✕

Name: IntProp-1

Contact Property Options

| Tangential Behavior |
| Normal Behavior |
| **Heat Generation** |

Mechanical   Thermal   Electrical   ✎

Heat Generation

Fraction of dissipated energy caused by friction or electric
currents that is converted to heat:

    ◯ Use default (1.0)

    ⦿ Specify: 9.9

Fraction of converted heat distributed to slave surface:

    ◯ Use default (0.5)

    ⦿ Specify: 0.99

          OK   Cancel

Fig. 4.67 Contact surface

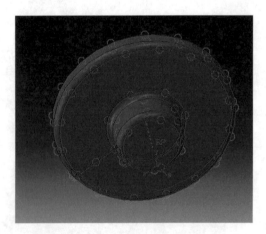

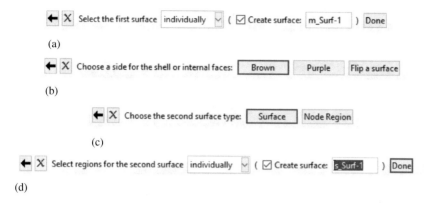

(a)

(b)

(c)

(d)

Fig. 4.68 Approving the selected region

Fig. 4.69 Selected region in the workpiece

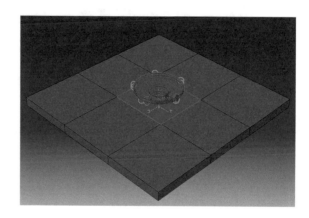

Fig. 4.70 Approving the region

Fig. 4.71 Confirming the interaction part

4.6 Loads and the Boundary Conditions

Firstly, we should fix the sides and the bottom surfaces of the workpiece.

BC → Create → Continue → Select sides and bottom surfaces → Done → Select Encastre → OK (Figs. 4.72, 4.73, 4.74 and 4.75).

The next step is selecting velocity boundary conditions:

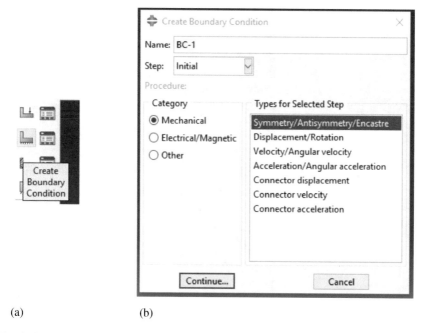

(a)   (b)

Fig. 4.72 Creating the boundary condition

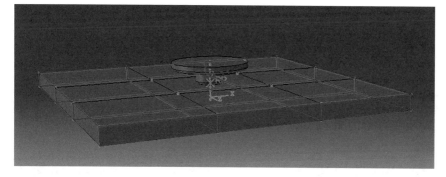

Fig. 4.73 Selecting the region

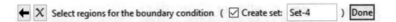

Fig. 4.74 Approving the region

Fig. 4.75 Applying the boundary condition

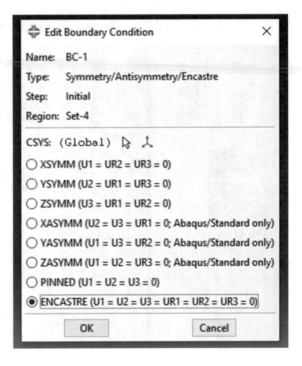

BC → Create → Select velocity / angular velocity → Continue → Select the tool reference point → Done → Close all velocity boundary conditions for the tool → OK (Figs. 4.76, 4.77 and 4.78).

The next step is editing the velocity boundary conditions for step one.

BC → Manage select → BC-2 Step 1 → Type (-25) radian per second for VR2, Note that V2 should be free in this step (Figs. 4.79 and 4.80).

The next step is defining the tool penetration depth.

BC → Create → Select displacement/Rotation → Continue → Select the tool reference point → Type (-0.005) for U2 → Select create amplitude → Tabular → Continue → Apply amplitude data → OK specify → APM 1 for this step → OK (Fig. 4.81, 4.82, 4.83 and 4.84).

The next step is to create a predefined temperature.

Predefined field → Create → Select other, temperature → Select SETS from the bottom right side → Select the entire workpiece sets → Continue → Type 25 for the magnitude → OK (Figs. 4.85, 4.86, 4.87 and 4.88).

After that the initial temperature for the tool should be defined.

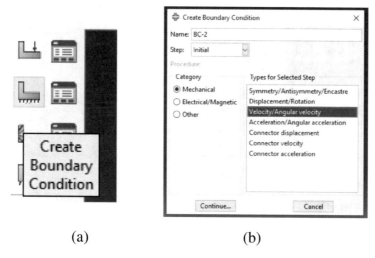

(a)   (b)

Fig. 4.76 Creating the boundary condition

Fig. 4.77 Confirming the boundary condition

Fig. 4.78 Selecting the
limitations of the boundary
conditions

Predefined field → Create → Select other, temperature → Select SETS from
the bottom right side → Select the entire tool sets → Continue → Type 25 for the
magnitude → OK (Figs. 4.89, 4.90, 4.91 and 4.92).

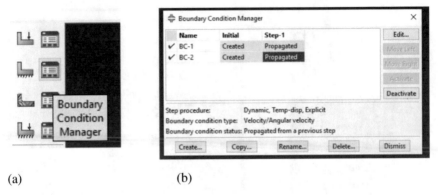

(a)   (b)

Fig. 4.79 Applying the boundary condition

Fig. 4.80 Velocity boundary condition

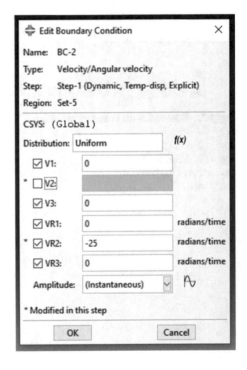

4.7   Meshing

Firstly we have to select the mesh for the workpiece.

Seed → Instance → Select the workpiece → Click on the middle mouse button (scroll wheel) or Done → Modify the mesh size to (0.005) → OK → Done (Figs. 4.93, 4.94, 4.95 and 4.96).

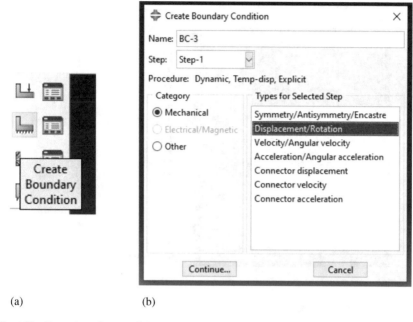

(a)   (b)

Fig. 4.81 Create boundary condition

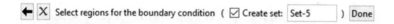

Fig. 4.82 Selecting the region

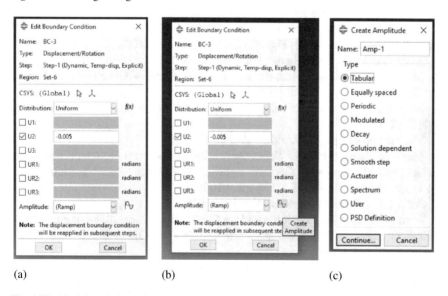

(a)   (b)   (c)

Fig. 4.83 Applying the boundary condition

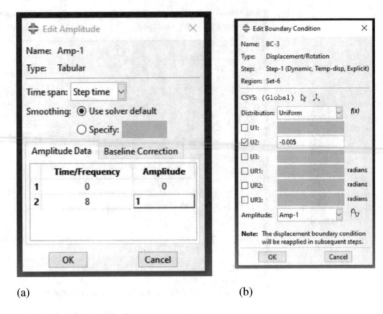

(a)   (b)

Fig. 4.84 Applying the amplitude

After that for the welding area smaller size of the mesh (in comparison with the default size) should be considered [17, 18]. Note that if you select smaller mesh size, your accuracy would increase. In contrast, the computational time will be increased [6, 19, 20].

Select the Seed → Edges → Select the center edge of the workpiece → Click on the middle mouse button (scroll wheel) → Done → Modify the mesh size to (0.003 or smaller) → OK → Done (Figs. 4.97, 4.98, 4.99 and 4.100).

After that the mesh should be applied to the workpiece.

Mesh → Instance → Select the workpiece → Done → Done (Figs. 4.101 and 4.102).

Note that the users can select smaller mesh size to increase the accuracy [12, 21, 22].

Then the element type should be assigned.

Mesh → Element type → Select the workpiece → Done → Select explicit → Select coupled temperature displacement → Select the reduced integration → OK → Done (Figs. 4.103, 4.104, 4.105 and 4.106).

After that the mesh for the tool should be selected and assigned.

Seed → Instance → Select the tool → Click on the middle mouse button (scroll wheel) or Done → Modify the mesh size to (0.002) → OK → Done (Figs. 4.107, 4.108, 4.109 and 4.110).

After that the mesh should be applied to the tool.

Mesh → Instance → Select the tool → Done → Done (Figs. 4.111 and 4.112).

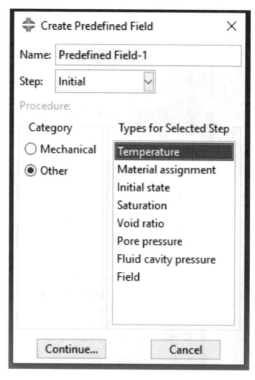

(a) (b)

Fig. 4.85 Creating the predefined field

Fig. 4.86 Selecting the set

Again note that, users can select smaller mesh size to increase the accuracy [12, 21, 22].

Then, the element type should be assigned.

Mesh → Element type → Select the tool set (from the bottom right side of the page) → Select the assigned set for the tool → Continue → Select explicit → OK → Dismiss → Done (Figs. 4.113, 4.114, 4.115, 4.116, 4.117 and 4.118).

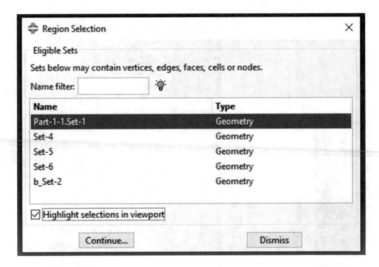

Fig. 4.87 Selecting the set

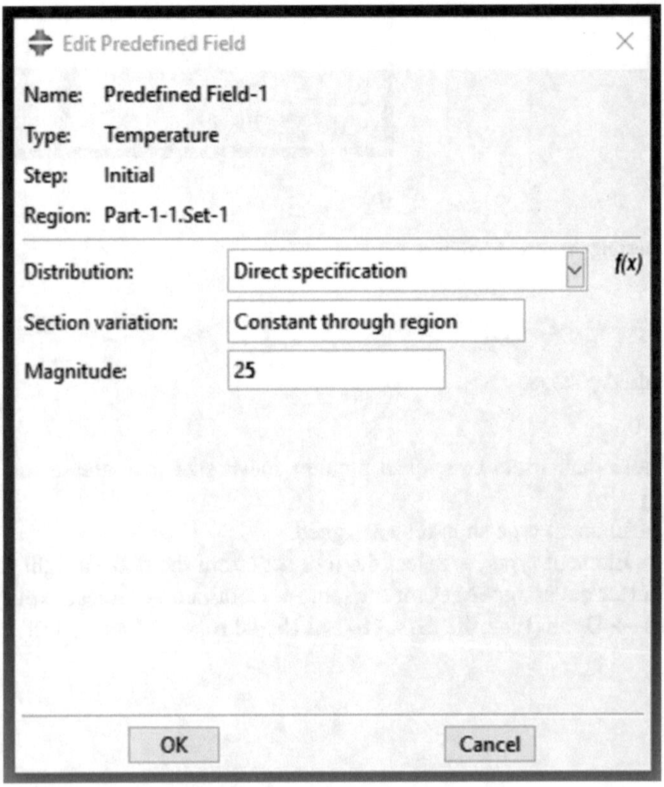

Fig. 4.88 Applying the value of 25

Fig. 4.89 Creating the
predefined field

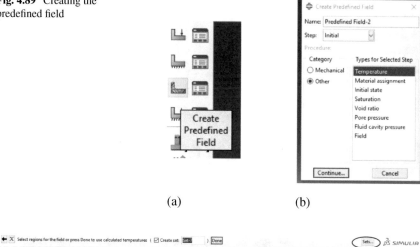

(a)   (b)

Select regions for the field or press Done to use calculated temperatures (☑ Create set: Set-7) Done   Sets...   ⫸ SIMULIA

Fig. 4.90 Selecting the set

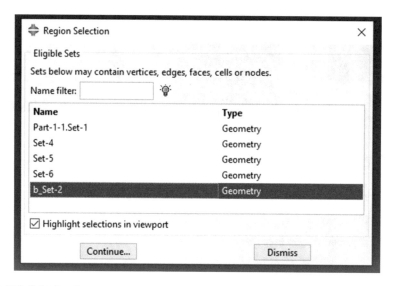

Fig. 4.91 Selecting the set

Fig. 4.92 Applying the value of 25

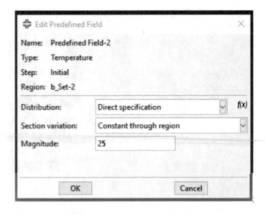

(a)

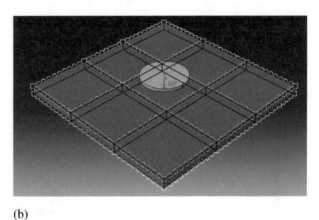

(b)

Fig. 4.93 Selecting the part

Select the part instances to be assigned global seeds   Done

Fig. 4.94 Applying the part instances

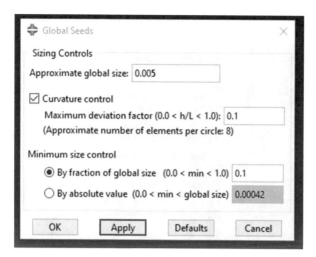

Fig. 4.95 Global seeds selection

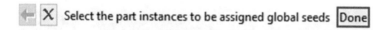

Fig. 4.96 Selecting the part instances

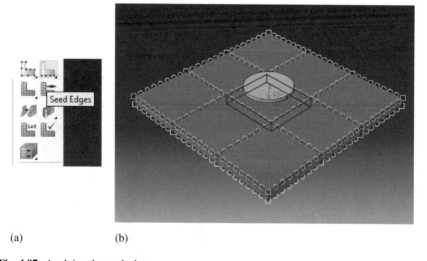

(a)   (b)

Fig. 4.97 Applying the seed edges

Fig. 4.98 Selecting the assigned region

Fig. 4.99 Selecting the local
seeds

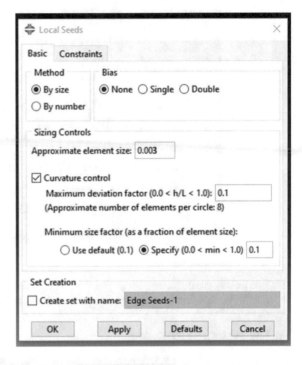

Fig. 4.100 Assigning the region

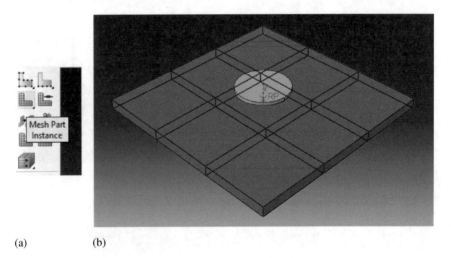

(a)   (b)

Fig. 4.101 Selecting the region

Fig. 4.102 Applying the part instances

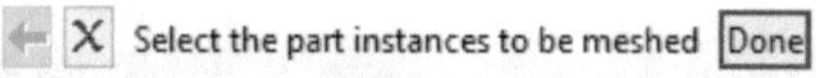

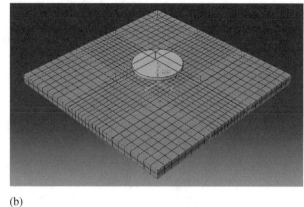

Select the part instances to be meshed Done

(a)

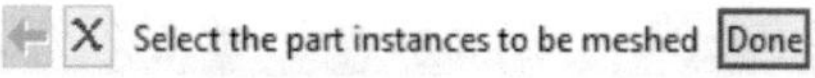

Select the part instances to be meshed Done

(b)

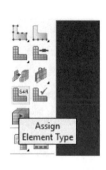

(a)

(b)

Fig. 4.103 Assigning the element type for the workpiece

Fig. 4.104 Selecting the region

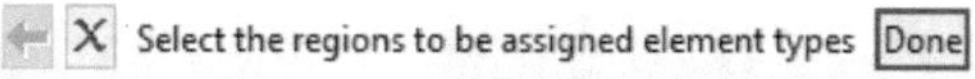

Select the regions to be assigned element types Done

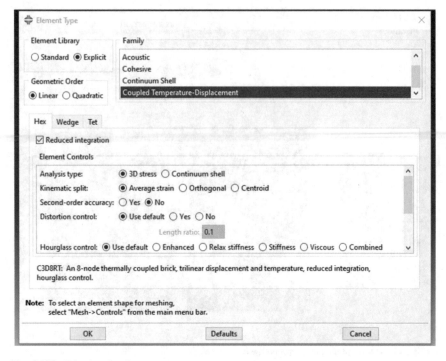

Fig. 4.105 Selecting the element type

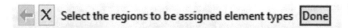

Fig. 4.106 Applying the region

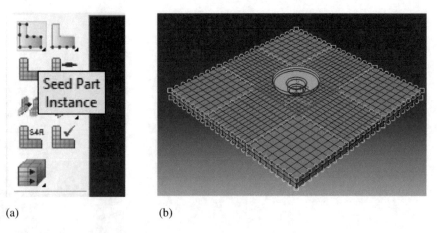

(a)   (b)

Fig. 4.107 Seed part instances

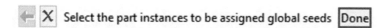

Fig. 4.108 Approving the region

Fig. 4.109 Selecting the size of the mesh

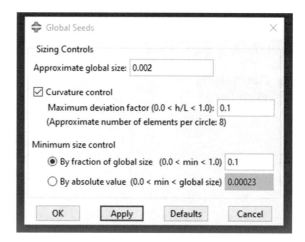

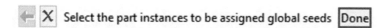

Fig. 4.110 Selecting the part instances

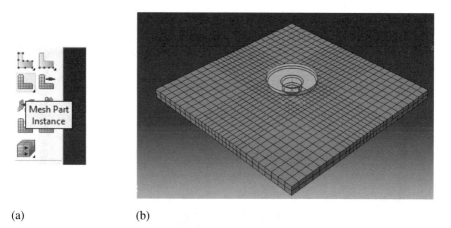

(a)   (b)

Fig. 4.111 Mesh part instances

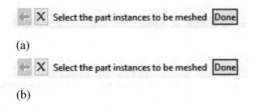

(a)

(b)

Fig. 4.112 Approving the region

Fig. 4.113 Assigning the
element type for the tool

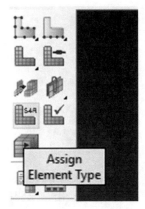

Fig. 4.114 Selecting the set

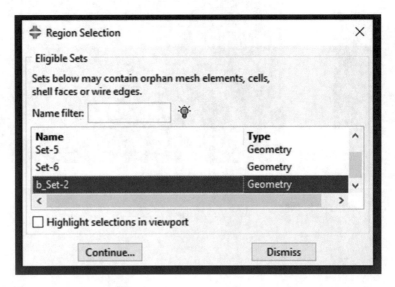

Fig. 4.115 Selecting the set between sets

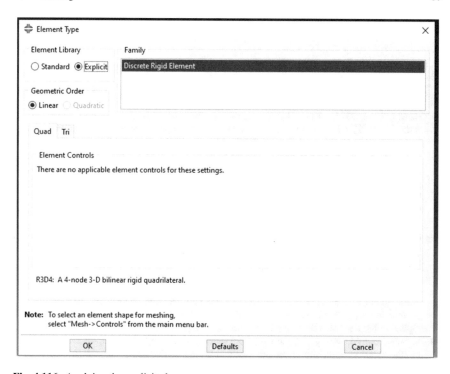

Fig. 4.116 Applying the explicit element type

Fig. 4.117 Selecting the set

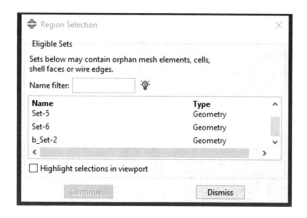

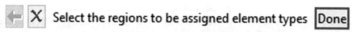

Fig. 4.118 Confirming the region

References

1. M. Awang, *The Advances in Joining Technology* (Springer, 2019)
2. B. Meyghani, M. Awang, Developing a finite element model for thermal analysis of friction stir welding (FSW) using hyperworks, in *Advances in Material Sciences and Engineering* (Springer, 2020), pp. 619–628
3. B. Meyghani, M.B. Awang, Prediction of the temperature distribution during friction stir welding (FSW) with a complex curved welding seam: application in the automotive industry, in *MATEC Web of Conferences*, vol. 225 (EDP Sciences, 2018), p. 01001
4. B. Meyghani, M. Awang, A comparison between the flat and the curved friction stir welding (FSW) thermomechanical behaviour, *Archives of Computational Methods in Engineering,* pp. 1–14 (2019)
5. B. Meyghani, M.B. Awang, M. Momeni, M. Rynkovskaya, Development of a finite element model for thermal analysis of friction stir welding (FSW), in *IOP Conference Series: Materials Science and Engineering*, vol. 495, no. 1 (IOP Publishing, 2019), p. 012101
6. B. Meyghani, M.B. Awang, R.G.M. Poshteh, M. Momeni, S. Kakooei, Z. Hamdi, The effect of friction coefficient in thermal analysis of friction stir welding (FSW), in *IOP Conference Series: Materials Science and Engineering*, vol. 495, no. 1 (IOP Publishing, 2019), p. 012102
7. B. Meyghani, M. Awang, S. Emamian, A comparative study of finite element analysis for friction stir welding application. ARPN J. Eng. Appl. Sci., **11**(22), 12984–12989 (2016)
8. B. Meyghani, M. Awang, S. Emamian, A mathematical formulatrion for calculating temperature dependent friction coefficient values: application in friction stir welding (FSW), in *Defect and Diffusion Forum*, vol. 379 (Trans Tech Publ., 2017), pp. 73–82
9. B. Meyghani, M. Awang, S. Emamian, "Introducing an enhanced friction model for developing inertia welding simulation: a computational solid mechanics approach," (in en). Int. J. Eng. **34**(3), 737–743 (2021)
10. B. Meyghani, M. Awang, C. Wu, Thermal analysis of friction stir welding with a complex curved welding seam. Int. J. Eng. **32**(10), 1480–1484 (2019)
11. B. Meyghani, S. Emamian, M. Awang, C.S. Wu, Finite element modeling of nano porous sintered silver material (Springer Singapore, Singapore, 2020), pp. 55–67
12. B. Meyghani, M. Awang, C. Wu, Finite element modelling of friction stir welding (FSW) on a complex curved plate. J. Adv. Joining Process. 100007 (2020)
13. B. Meyghani, M. Awang, C. Wu, Probabilistic finite element analysis of the deflection on a beam, in *IOP Conference Series: Materials Science and Engineering*, vol. 863, no. 1 (IOP Publishing, 2020), p. 012002
14. B. Meyghani, Thermomechanical analysis of friction stir welding (FSW) on curved plates by adapting calculated temperature dependent properties (Universiti Teknologi Petronas, 2018)
15. B. Meyghani, M. Awang, S. Emamian, N.M. Khalid, Developing a finite element model for thermal analysis of friction stir welding by calculating temperature dependent friction coefficient, in *2nd International Conference on Mechanical, Manufacturing and Process Plant Engineering*, pp. 107–126 (Springer, 2017)
16. B. Meyghani, M. Awang, S. Emamian, M.K.B.M. Nor, Thermal modelling of friction stir welding (FSW) using calculated Young's modulus values," in *The Advances in Joining Technology* (Springer, 2019), pp. 1–13
17. B. Meyghani, M.B. Awang, Prediction of the temperature distribution during friction stir welding (FSW) with a complex curved welding seam: application in the automotive industry. MATEC Web Conf. **225**, 01001 (2018)
18. B. Meyghani, M.B. Awang, S.S. Emamian, M.K.B. Mohd Nor, S.R. Pedapati, A comparison of different finite element methods in the thermal analysis of friction stir welding (FSW). Metals **7**(10), 450 (2017)
19. B. Meyghani, C. Wu, Progress in thermomechanical analysis of friction stir welding. Chin. J. Mech. Eng. **33**(1), 12 (2020)

20. B. Meyghani, M. Awang, A novel tool path strategy for modelling complicated perpendicular curved movements, in *Key Engineering Materials*, vol. 796 (Trans Tech Publ., 2019), pp. 164–174
21. B. Meyghani, M. Awang, P. Bokam, B. Plank, C. Heinzl, K. Siow, Finite element modeling of nano porous sintered silver material using computed tomography images. Materialwiss. Werkstofftech. **50**(5), 533–538 (2019)
22. B. Meyghani, M. Awang, C. Wu, Thermal analysis of friction stir processing (FSP) using arbitrary Lagrangian-Eulerian (ALE) and smoothed particle hydrodynamics (SPH) meshing techniques. Materialwiss. Werkstofftech. **51**(5), 550–557 (2020)

Chapter 5
Post Processor and Visualization of the Results

5.1 Post Processer

In the first step of the postprocessor module, a job should be created.

Job → Create → Continue → OK (Fig. 5.1).

Then, the run should be started.

Job → Manager → Submit → Waiting until completing the job → After completing the job click on results (Figs. 5.2, 5.3 and 5.4).

In this module you can visual all of the results (for example stress, temperature, etc.) (Figs. 5.5, 5.6 and 5.7).

© The Author(s), under exclusive license to Springer Nature Singapore Pte Ltd. 2022
B. Meyghani and M. Awang, *Welding Simulations Using ABAQUS*,
https://doi.org/10.1007/978-981-19-1320-4_5

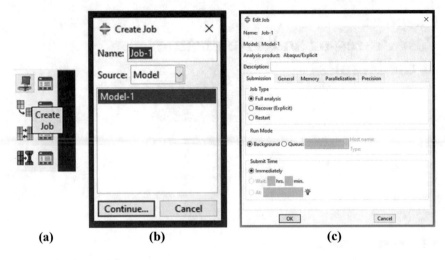

Fig. 5.1 Creating the job

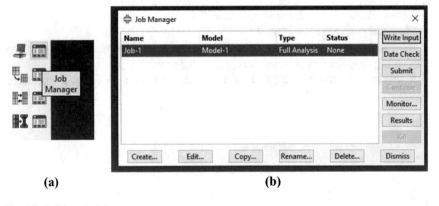

Fig. 5.2 Editing the job manager

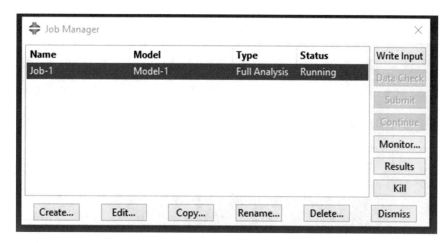

Fig. 5.3 Running the job

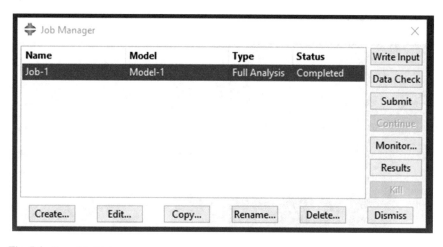

Fig. 5.4 Completed job

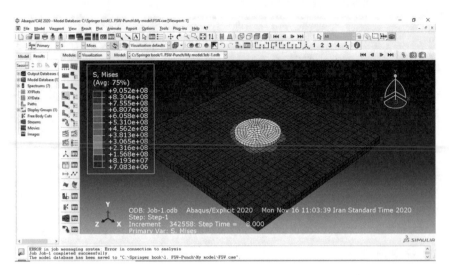

Fig. 5.5 Stress across the plate

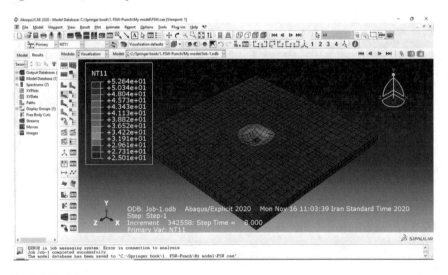

Fig. 5.6 Nodal temperature

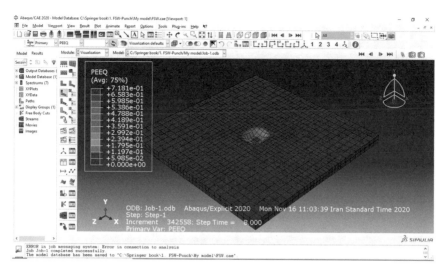

Fig. 5.7   Equivalent plastic strain

Chapter 6
Thermomechanical Simulation of FSW Using User Defined Subroutine Modeling Technique

6.1 Preparation of the Model

It should be noted that modeling FSW by heat flux method should be done into two steps. The first one is thermal step and the second one is the mechanical step. Thus, two models should be defined.

As mentioned earlier the first step is creating a name for model one (Thermal model).

Model → Manager → Rename → Thermal model → OK (Fig. 6.1).

After that the part should be created (it should be noted that in this method, only the workpiece should be simulated and there is no need to create the tool). In addition, since the workpiece is symmetrical, we can only create half part of it and the second half can be added in the Job module. This issue decrease the computational costs of the simulation.

Part → Create a 3D, deformable, solid extrusion model (Fig. 6.2).

Create rectangle → at the bottom part the rectangle points should be applied (Figs. 6.3, 6.4, 6.5 and 6.6).

After that middle mouse button (scroll wheel) should be clicked two times.

Then, the thickness of 6 mm is selected for workpiece.

6.2 Material Property

In this module, the material property should be applied to the module.

Material → Create (Figs. 6.7 and 6.8).

Since in the previous simulation, we have completely explained the material property, in this section only the values of the required material property are shown in different tables (including conductivity, density, elastic, expansion, plastic and specific heat).

The next step is creating a solid homogenous section.

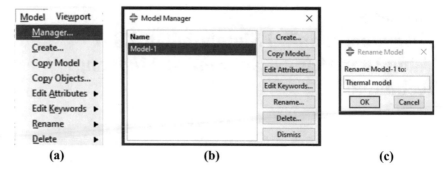

Fig. 6.1   Creating the new model

Fig. 6.2   Creating the part

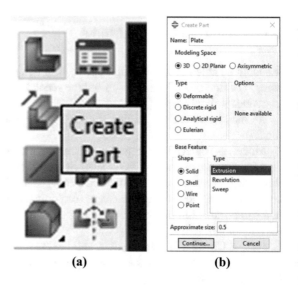

Section → Create (Fig. 6.9).

The next step is assigning this section.

Assign → Section → Selecting the workpiece → Done → OK (Figs. 6.10, 6.11 and 6.12).

6.3   Assembly of the Model

Instance → Create → Selecting bodies → OK (Figs. 6.13 and 6.14).

Then, there is a need to rotate the model.

Instance → Rotate → Select the back side point of the model → The middle mouse button (scroll wheel) should be clicked one time, then the angle should be

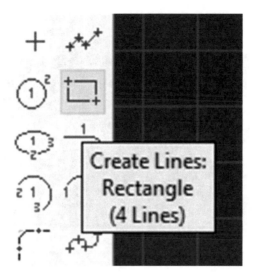

Fig. 6.3 Creating the rectangle

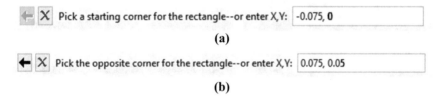

(a)

(b)

Fig. 6.4 Rectangle dimentions

Fig. 6.5 Extrusion values

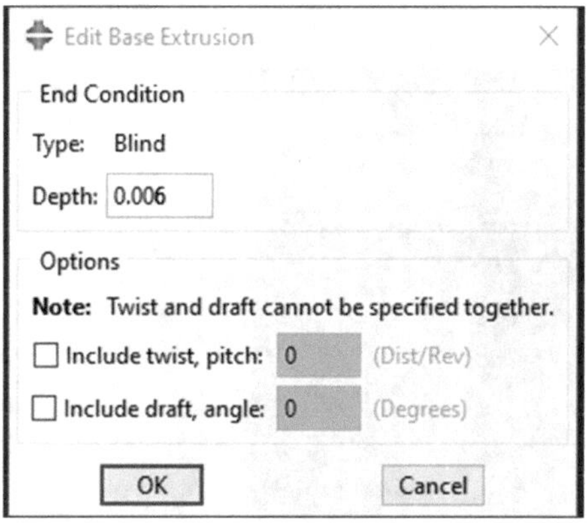

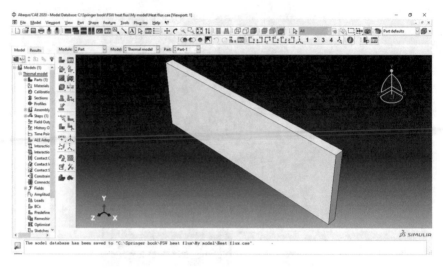

Fig. 6.6 Final part

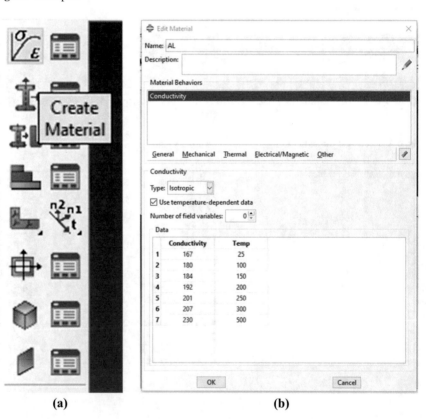

(a)   (b)

Fig. 6.7 Creating the material property

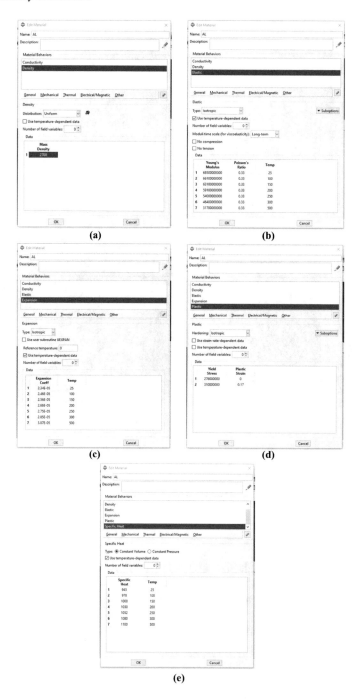

Fig. 6.8 Material property definition

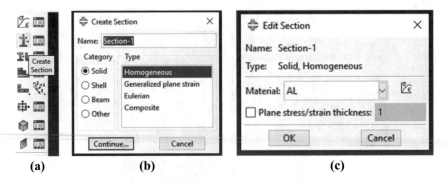

Fig. 6.9 Creating the solid homogenous section

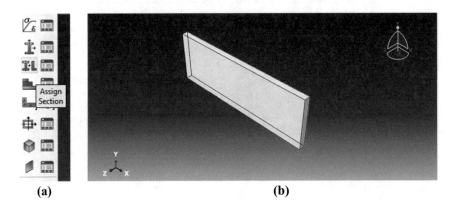

Fig. 6.10 Assigning the section

Fig. 6.11 Confirming the region

checked (90°) → The middle mouse button (scroll wheel) should be clicked one more time → OK (Figs. 6.15, 6.16, 6.17 and 6.18).

After that, the center of the coordinate system should be changed.

Instance → Translate → Select the starting point → Select the center of the coordinate system as the end point → OK (Figs. 6.19, 6.20, 6.21 and 6.22).

Fig. 6.12 Editing the
section assignment

Fig. 6.13 Creating the
instance part

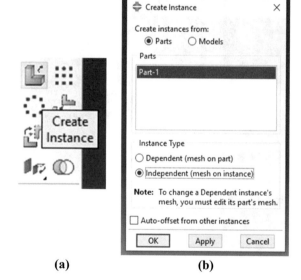

(a)   (b)

6.4   Simulation Step Creation

In this section two steps are defined for the welding and the cooling.

   Step → Create → Selecting a heat transfer step → Continue (Fig. 6.23).

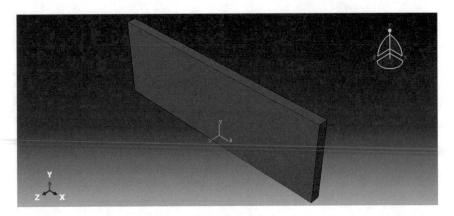

Fig. 6.14 The created part

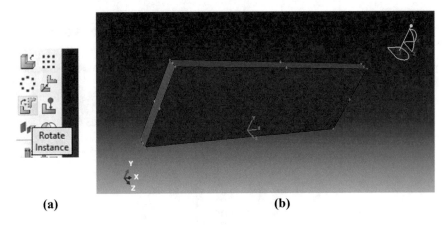

(a)   **(b)**

Fig. 6.15 Selection the rotation point

Fig. 6.16 The rotation angle

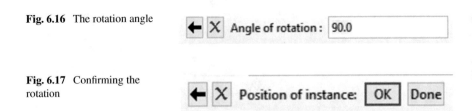

Fig. 6.17 Confirming the rotation

- Here it should be noted that the welding transverse speed is selected to be 60 m/min. Since the workpiece length is 150 mm, therefore the welding step time should be 150 s.
- It is also recommended to use the FIX instead of automatic in the incrementation tab.

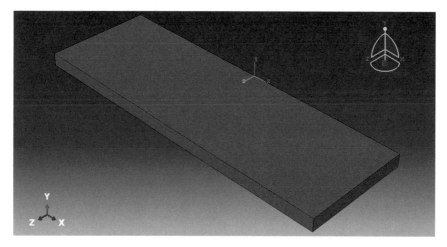

Fig. 6.18 Positioning

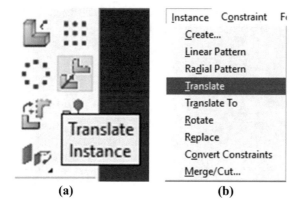

(a)   **(b)**

Fig. 6.19 Creating the instance

- Maximum number of increments should be a large number.
- For the increment size, the literature recommended the use of 0.25 [1–3].

  Then OK.
  The next step is the cooling step.
  In this section two steps are defined for the welding and the cooling.
  Step → Create → Selecting a heat transfer step → Continue (Fig. 6.24).

- 300 s is considered for the cooling step.
- In this step it is recommended to use automatic in the incrementation tab.
- Maximum number of increments should be a large number.
- The increment size should be varied between 0.25 and 25.
- Maximum allowed temperature for each increment also should be 100.

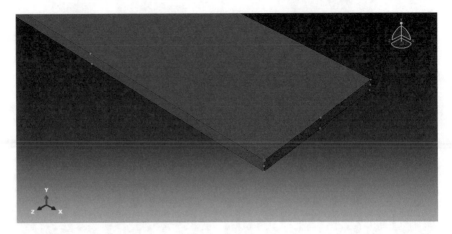

Fig. 6.20 Selecting the point

Fig. 6.21 Selecting the point for the coordinate system

Fig. 6.22 Confirming the
position of the instance

Then OK.

Then, the output for the last step should be modified and the temperature for elements should be added.

Output → Field output requests → Manager → Click edit to modify it for the welding step → In the thermal section add TEMP output → OK (Figs. 6.25 and 6.26).

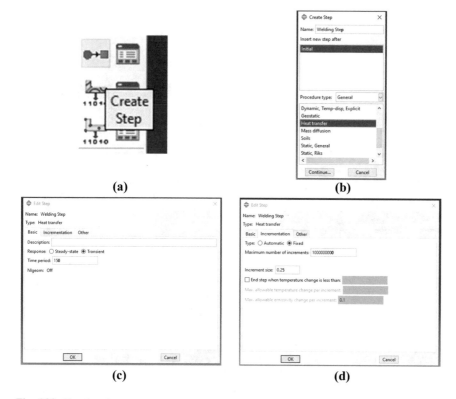

Fig. 6.23 Creating the step

Since we have to use the results of the thermal step for the mechanical step, we have to request the software to restart the job after completing.

Output → Restart request → Select number 1 for frequency (Fig. 6.27).

6.5   Interaction Definitions

In the first section of this module the absolute zero temperature and the Stefan-Boltzmann constant should be defined for the model.

Model → Edit attributes (Fig. 6.28).

Then, a boundary condition for the bottom surface of the workpiece should be selected.

Interaction → Create select surface film condition → Continue → Select the bottom surface of the workpiece → After that, the middle mouse button (scroll wheel) should be clicked two times → Edit the film coefficient and the sink temperature → OK (Figs. 6.29, 6.30 and 6.31).

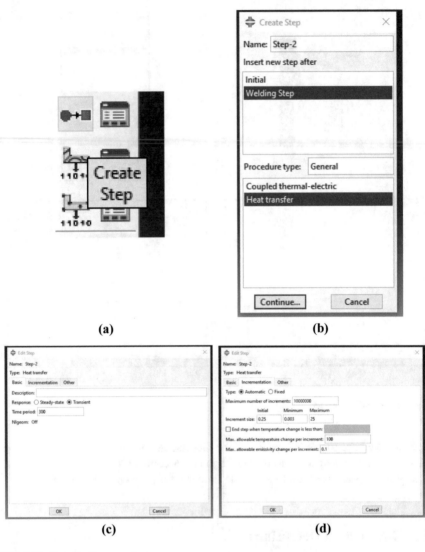

Fig. 6.24   Creating the second step

For the other remaining surfaces we need to consider a boundary condition between the heat and air.

Interaction → Create select surface film condition → Continue → Select the other remaining surfaces of the workpiece (except the welding seam) → After that middle mouse button (scroll wheel) should be clicked two times → Edit the film coefficient and the sink temperature → OK (Figs. 6.32, 6.33 and 6.34).

The next step is to create the surface radiation.

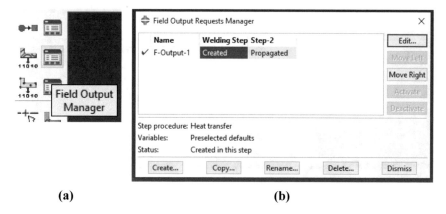

(a)   (b)

Fig. 6.25 Requesting the filed output

Fig. 6.26 Requesting the filed output

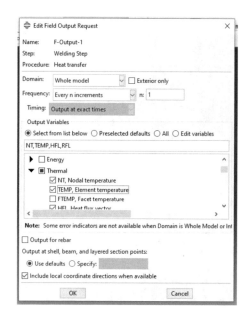

Interaction → Create select radiation → Continue → Select the 4 edges of the workpiece (except the welding seam) → After that, the middle mouse button (scroll wheel) should be clicked two times → Edit the emissivity and the ambient temperature → OK (Figs. 6.35, 6.36 and 6.37).

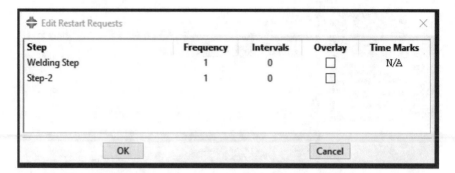

Fig. 6.27 Requesting the restart

Fig. 6.28 Defining the
interaction

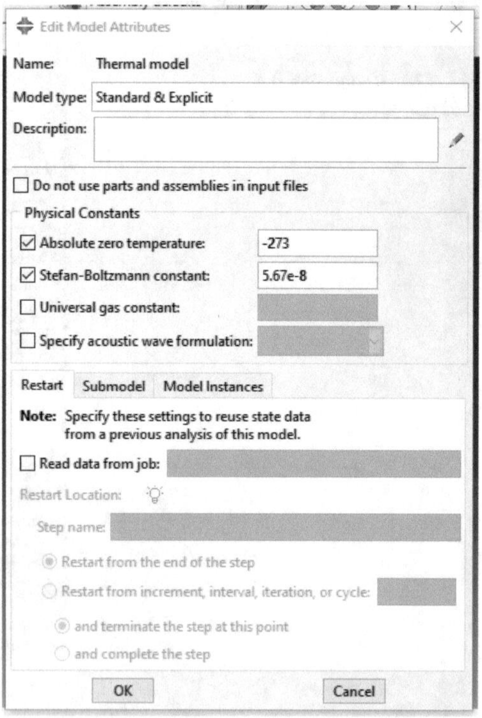

6.6 Mesh Module

In this example we used another method for creating the mesh.

Seed → Edges → Selecting the thickness of the workpiece → Done → Selecting
by number Type 6 → OK (Fig. 6.38).

Now, we have to set the elements for the length of the workpiece.

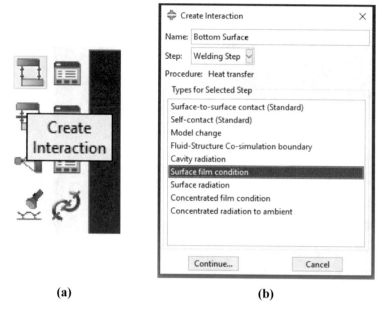

(a)   (b)

Fig. 6.29 Defining the surface film condition

Fig. 6.30 Selecting the surface

Fig. 6.31 Editing the interaction

(a)

(b)

Fig. 6.32 Creating the interaction

Fig. 6.33 Selecting the surface

Fig. 6.34 Editing the interaction

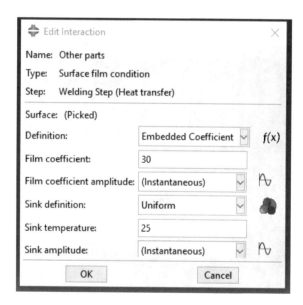

Seed → Edges → Selecting the length of the workpiece → Done → Selecting by number Type 150 → OK (Fig. 6.39).

For the width of the workpiece we have to use another method, because we have to apply a fine mesh for the welding area. In contrast, those regions which are located far from the welding seam should have a coarse mesh.

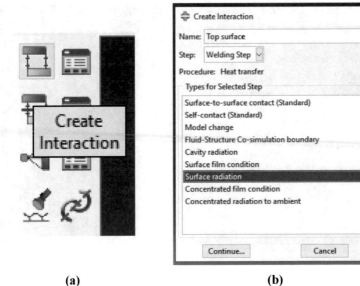

(a)   (b)

Fig. 6.35 Creating the interaction

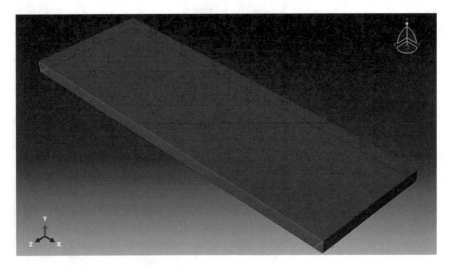

Fig. 6.36 Selecting the surface

Seed → Edges → Selecting the width of the workpiece → Done → Selecting single → Flip the mesh direction → Type 25 and 5 for the number of the elements and bias ratio, respectively → OK → Done (Fig. 6.40).

Now, the element type should be selected.

Mesh → Element type → Select heat transfer → Ok (Fig. 6.41).

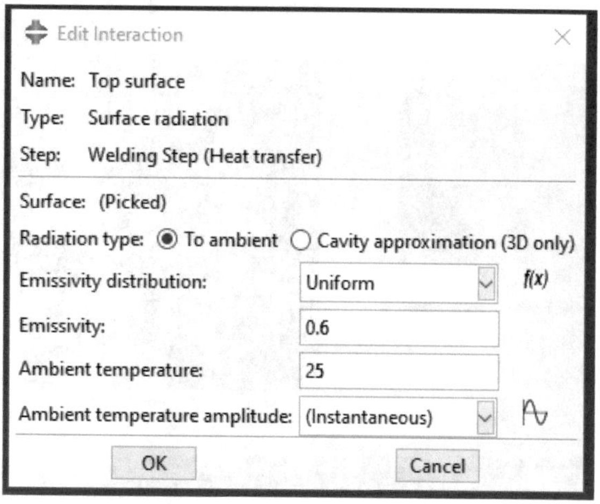

Fig. 6.37 Editing the
interaction

In this step the mesh should be applied to the model.

Mesh Instance → Select the Workpiece → Yes (Fig. 6.42).

In the load module the initial conditions and the subroutine should be applied.

Predefined field → Create (Fig. 6.43).

Select other and then select temperature (Fig. 6.44).

The workpiece should be selected (Fig. 6.45).

After that, Done should be selected (Fig. 6.46).

Then type 25 as the initial temperature (Fig. 6.47).

6.7 Loads and Boundary Conditions

Again we have to move to the load module in order to define two heat fluxes (for the shoulder we have to apply a heat flux on the surface and for the pin, a volume heat flux should be applied).

First the heat flux on the surface.

Load → Create load → Select surface heat flux for the welding step → Continue → Select the upper surface of the workpiece → Done → Select User Defined and type 0 for the magnitude → OK (Fig. 6.48).

Now the volumetric heat flux should be applied.

Load → Create load → Select surface body heat flux for the welding step → Continue → Select the workpiece → Done → Select User defined and type 0 for the magnitude → OK (Fig. 6.49).

Now in the cooling step we have to deactive the defined heat fluxes.

Load → Manager → Select the heat fluxes for step 2 → Click on deactivate → Dismiss (Fig. 6.50).

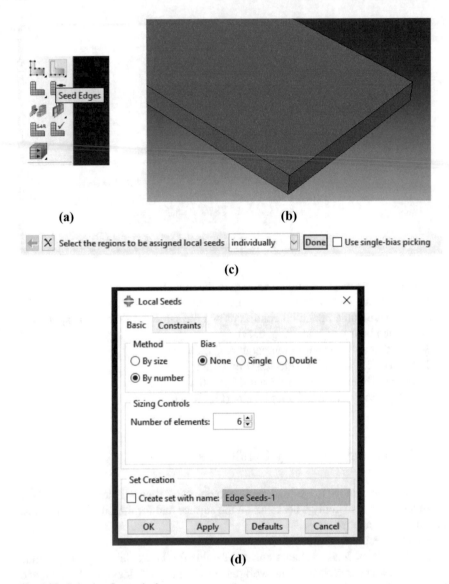

(a)   **(b)**

(c)

(d)

Fig. 6.38 Selecting the seed edge

The subroutine is a FORTRAN code. In order to have the subroutine please use the below text and save it as a FORTRAN code file (.for) in your note pad. The subroutine code is written in the box and other parts are extra descriptions in order to briefly explanation of the Subroutine.

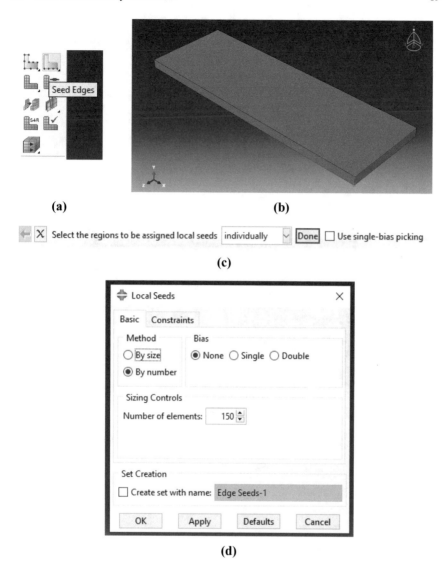

(a)

(b)

Select the regions to be assigned local seeds individually ⌄ Done ☐ Use single-bias picking

(c)

(d)

Fig. 6.39 Selecting the seed edge

Heading: Basic section of the subroutine (you can find it in the help part of the ABAQUS software).

- We have two heat fluxes then we should write two for the heat flux.
- We have to write the number 2 for the TIME, firstly for the step time and the second one is the whole simulation time.
- Three coordinates for X, Y and Z are defined.

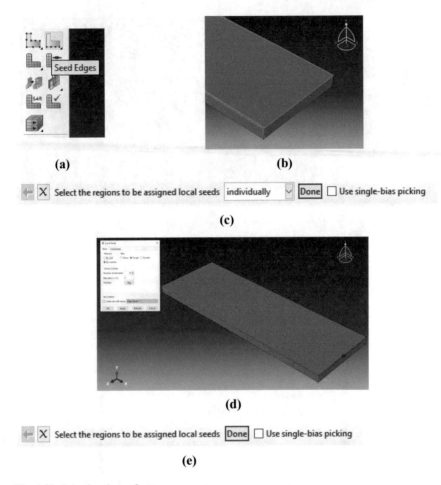

(a)   **(b)**

← X Select the regions to be assigned local seeds | individually ∨ | Done | ☐ Use single-bias picking

(c)

(d)

← X Select the regions to be assigned local seeds | Done | ☐ Use single-bias picking

(e)

Fig. 6.40 Selecting the surface

```
 1 JLTYP,TEMP,PRESS,SNAME)
SUBROUTINE DFLUX(FLUX,SOL,KSTEP,KINC,TIME,NOEL,NPT,COORDS,
C
INCLUDE 'ABA_PARAM.INC'
C
DIMENSION FLUX(2), TIME(2), COORDS(3)
CHARACTER*80 SNAME
T = Time(1)
```

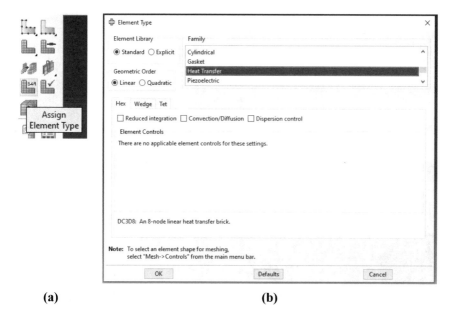

Fig. 6.41 Selecting the heat transfer element

Subroutine body: this part should be written by the user.

- Defining coordinates for X, Y and Z.
- Then, the starting point of the subroutine should be written. In our model because the coordinate system is located at the starting point we have select + 0.001. It means that our starting point is located exactly at the welding edges.
- Then, the welding transverse velocity (1 mm per second) should be defined.
- Then, the dynamic formulations for X, Y and Z should be written.
- After that, Xf, Yf and Zf should be defined as the location point for the heat flux at each time increment.
- "R" is the radius of the applied heat flux.
- R_p, R_s, and H_p are the pin radius, shoulder radius and height of the pin (the negative sign of the pin radius shows the direction in which the heat should be applied).
- After that, Flux for the pin and the shoulder are applied based on the reference [4].
- "JLTYP" shows the type of the flux (surface flux = when JLTYP is equal to 0 or volumetric flux = when JLTYP is equal to 1).
- It should be noted that, the area for the heat flux at the shoulder surface is defined as the difference between the shoulder diameter and the pin diameter.
- The volumetric heat flux is defined as the volume of the pin.

(a)

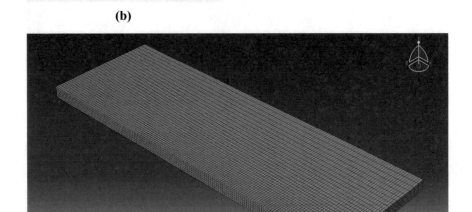

(c)

Fig. 6.42 Creating the mesh

Fig. 6.43 Creating the predefined field

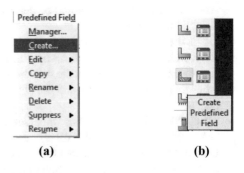

(a)   **(b)**

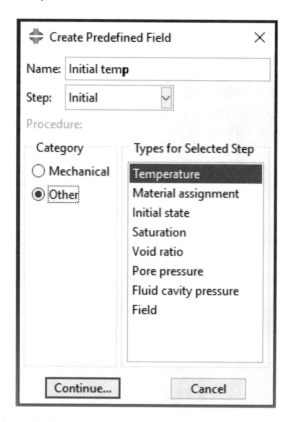

Fig. 6.44 Defining the initial temperature

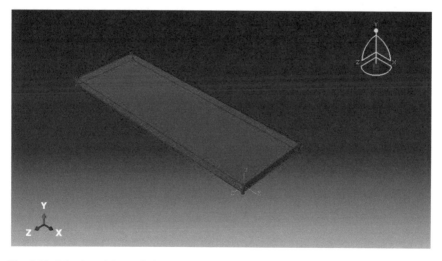

Fig. 6.45 Selecting of the workpiece

Fig. 6.46 Approving the selected workpiece

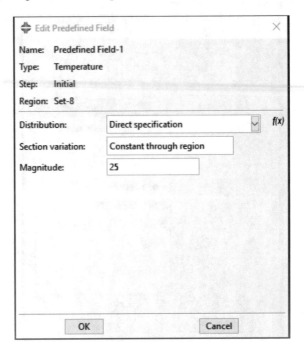

Fig. 6.47 Selecting the temperature

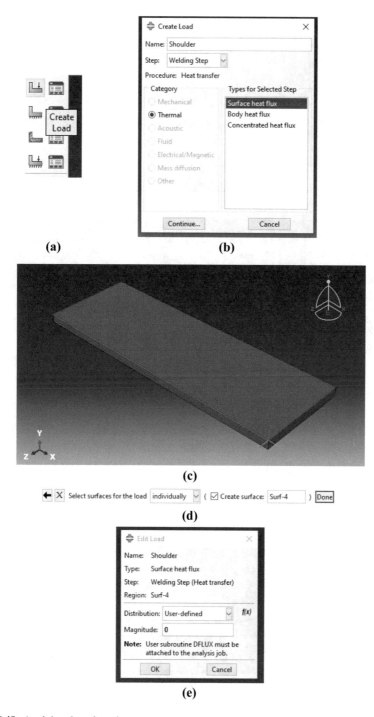

Fig. 6.48   Applying the subroutine

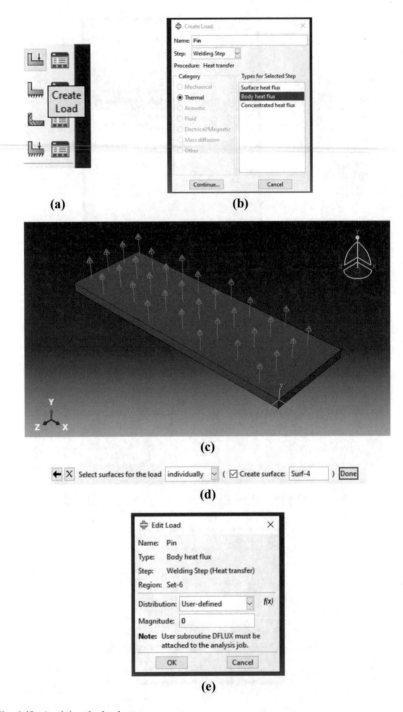

Fig. 6.49 Applying the load

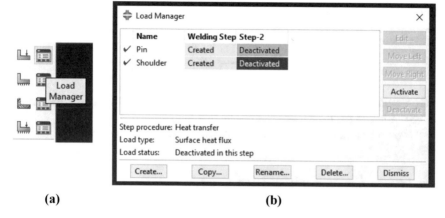

(a)   **(b)**

Fig. 6.50 Deactivation of the step

```
X = COORDS(1)
Y = COORDS(2)
Z = COORDS(3)
X0 = 0
Y0 = 0
Z0 = 0
Vx = -0.001
Vy = 0
Vz = 0
Xs = Vx*T + X0
Ys = Vy*T + Y0
Zs = Vz*T + Z0
Xf = X-Xs
Yf = Y-Ys
Zf = Z-Zs
R = SQRT(Xf**2 + Zf**2)
Rp = 0.003
Rs = 0.01
Hp = -0.0047
FluxPin = 3,794,640,264.77541
FluxShoulder = 1,194,333.62051282
IF(JLTYP.EQ.0.AND. T.GE.0.AND. R.LE.Rs.AND. R.GE.Rp)Then
FLUX(1) = FluxShoulder
ELSEIF(JLTYP.EQ.1.AND. T.GE.0.AND. R.LE.Rp.AND. Yf.GE.Hp)Then
Flux(1) = FluxPin
ENDIF
```

End of the subroutine.

RETURN
END

- It should be noted that the subroutine should be saved at the same folder of the FSW file and the format of the subroutine is (.for).
- It should be also considered that, the ABAQUS should be linked to Fortran.

6.8   Postprocessor (Thermal Model)

The last step is creating the job and run the thermal model.
Job → Create →Continue (Figs. 6.51 and 6.52).
In the general tab, the subroutine should be addressed.
Then click on OK (Fig. 6.53).
After that the job should be run.

Fig. 6.51 Creating the job

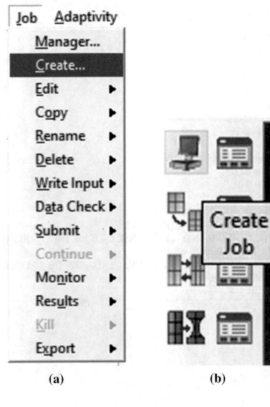

(a)   (b)

Fig. 6.52 Creating the
thermal model

Fig. 6.53 Defining the
subroutine for the simulation

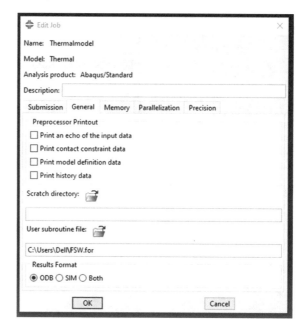

Job → Manager (Fig. 6.54).

Then, click on submit to run the simulation (Fig. 6.55).

Submit → Yes (Fig. 6.56).

As can be seen in the below figures the results for the thermal model can be seen.

Fig. 6.54 Managing the job

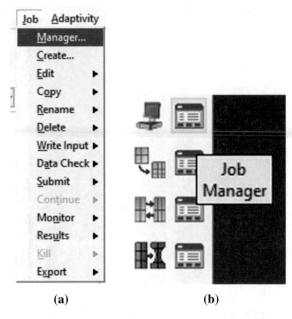

(a)   (b)

Fig. 6.55 Job manager part

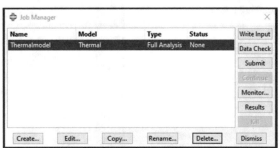

Fig. 6.56 Requesting to continue the job

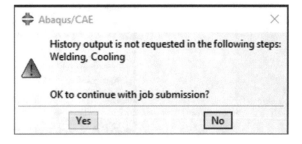

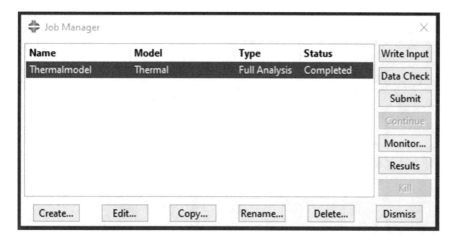

Fig. 6.57 Completed job

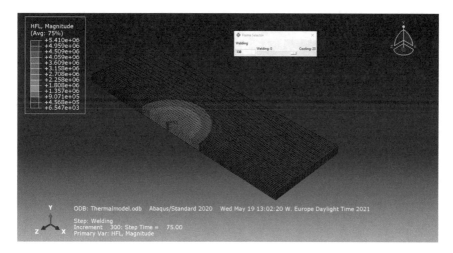

Fig. 6.58 Heat flux results

After completing the job, click on the results button to visual the results (Figs. 6.57, 6.58, 6.59 and 6.60).

6.9 Mechanical Step

After completing the thermal step the mechanical step should be defined for the simulation.

In the part module a copy from the previous thermal module should be created:

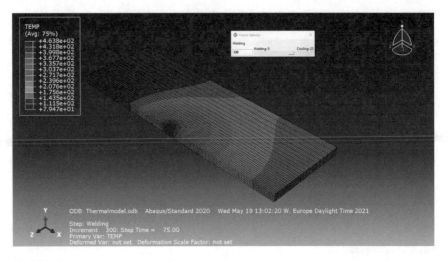

Fig. 6.59 Element temperature results

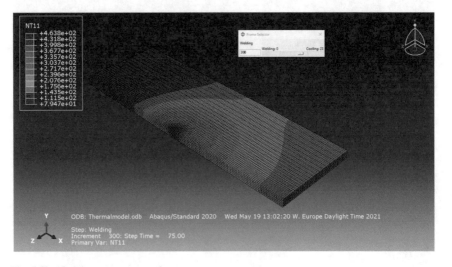

Fig. 6.60 Nodal temperature results

Model → Copy model → Thermal → Select the name → OK (Figs. 6.61 and 6.62).

Note: Please check and make sure that the mechanical model is selected in the task bar (Fig. 6.63).

Since the step and the type of analyze is different, the step module should be modified (Fig. 6.64).

Firstly, the cooling step should be deleted.

Delete → Yes (Fig. 6.65).

Fig. 6.61 Copying the model

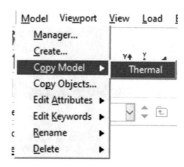

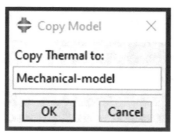

Fig. 6.62 Creating the mechanical model

Fig. 6.63 Checking the model

Fig. 6.64 Creating the step

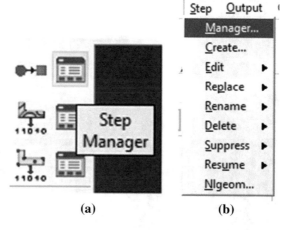

(a)   (b)

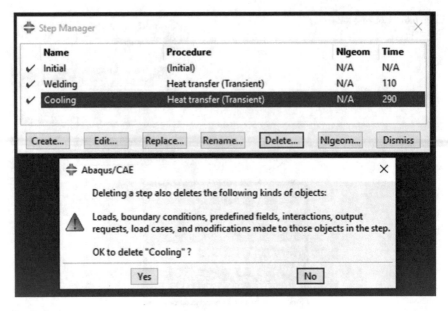

Fig. 6.65 Deleting the step

The welding step should be also renamed (Fig. 6.66):

Rename → Change the name → OK.

After that, it should be replaced with a static general step (Fig. 6.67).

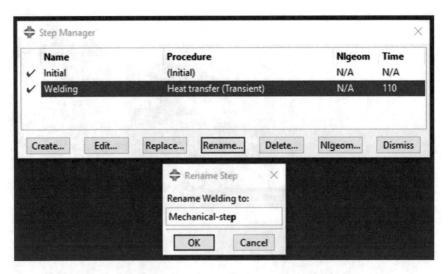

Fig. 6.66 Creating the mechanical step

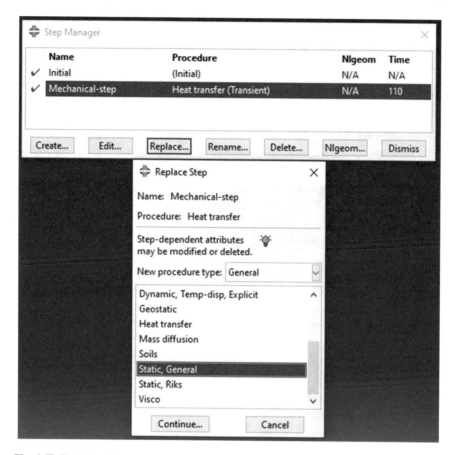

Fig. 6.67 Replacing the step

Replace → Select static general → Continue → Turn on the Nlgeom → Change the time period to 400 (the total amount of the steps in welding and cooling steps) → Modify the incrementation task bar (Figs. 6.68 and 6.69).

6.10 Loads and Boundary Conditions

Since the previously defined boundary conditions are no longer available, there is a need to define new boundary conditions.

BC Create → Select the initial step → Continue → Select the left surface → Done → Select Encastre → OK (Figs. 6.70, 6.71, 6.72, 6.73, 6.74 and 6.75).

Since the only the half part of the plate is modeled, there is a need to have an asymmetrical boundary condition.

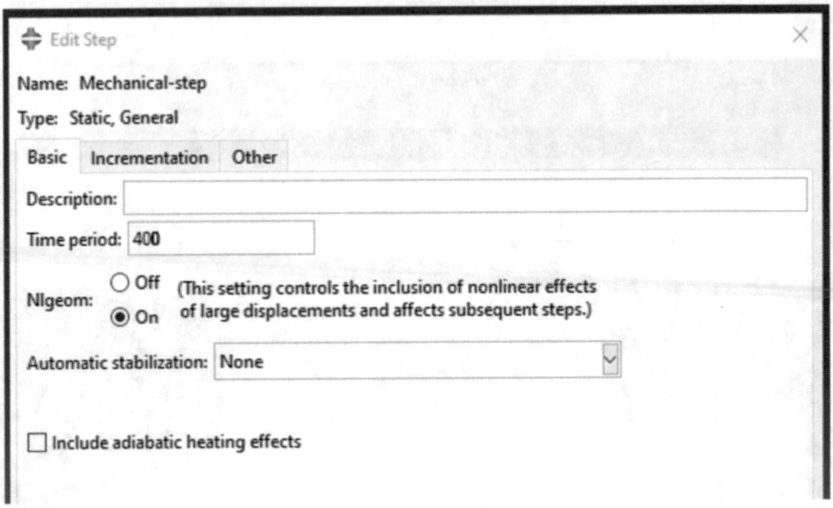

Fig. 6.68 Modifying the step

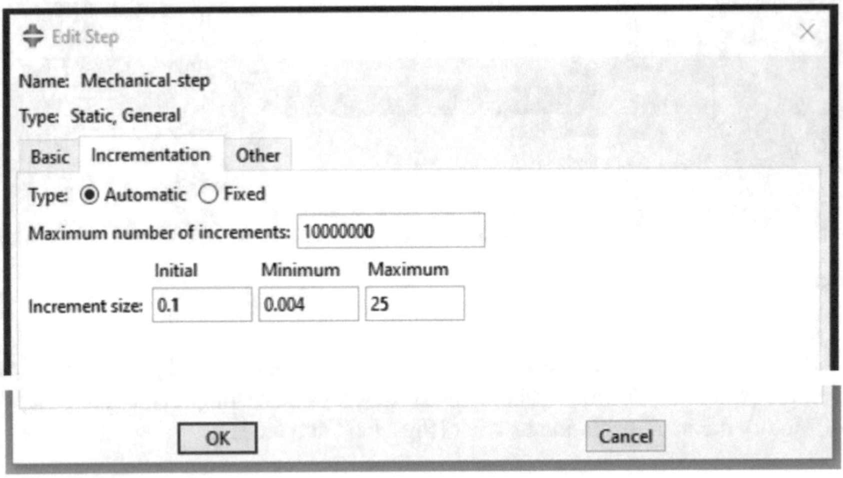

Fig. 6.69 Modifying the step

BC Create → Select the initial step → Continue → Select the right surface → one → Select ZSYMM → OK (Figs. 6.76, 6.77, 6.78, 6.79, 6.80 and 6.81).

After that, there is a need to add the results of the thermal model as a boundary condition for the mechanical model.

Create predefined field → Select the step module → Select other → Select temperature → Continue → Select the workpiece → Done → Select results or output database file → For the file name, the Odb file from the thermal step should be added

Fig. 6.70 Creating the boundary condition

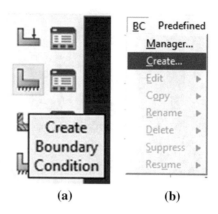

(a)   (b)

Fig. 6.71 Creating the boundary condition

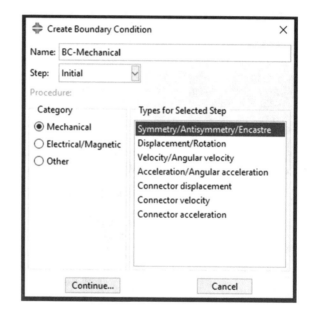

→ The beginning step and the beginning increment is 1 → The end increment is 464 (440 increments for step one, 24 increments for step 2, thus the total amount of 464 increments should be written) → OK (Figs. 6.82, 6.83, 6.84, 6.85 and 6.86).

6.11 Mesh Module

In this module, the element type should be changed.
Mesh → Element type → Select 3D stress (Figs. 6.87 and 6.88).

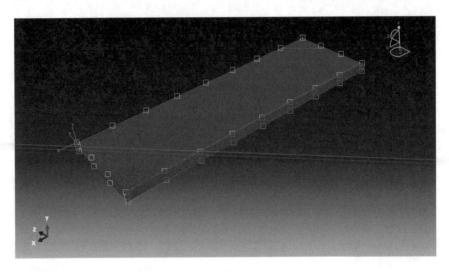

Fig. 6.72 Selecting the side of the workpiece

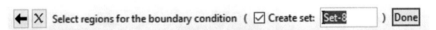

Fig. 6.73 Confirming the set

Fig. 6.74 Selecting the
boundary condition

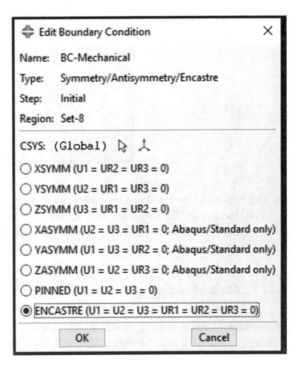

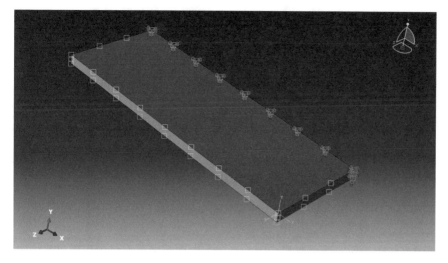

Fig. 6.75 Selecting the workpiece

Fig. 6.76 Creating the
boundary condition

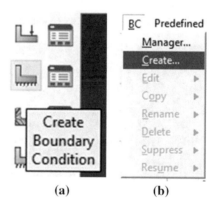

(a)   (b)

6.12   Returning to the Step Module in Order to Request the Results

Output → Field output request → Manager → Edit → Select preselected defaults
→ OK (Figs. 6.89, 6.90 and 6.91).

6.13   Postprocessor

Job → Create → Select the name (please double check in selecting the mechanical
model) → Continue → OK (Figs. 6.92, 6.93 and 6.94).

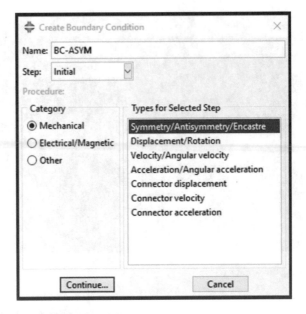

Fig. 6.77 Slecting the boundary condition

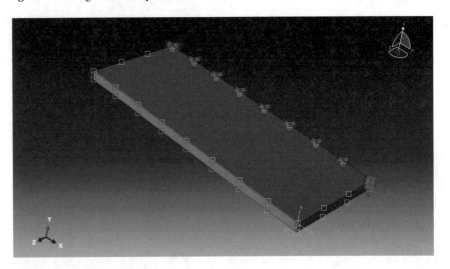

Fig. 6.78 Selecting the workpiece

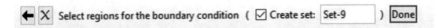

Fig. 6.79 Confirming the set

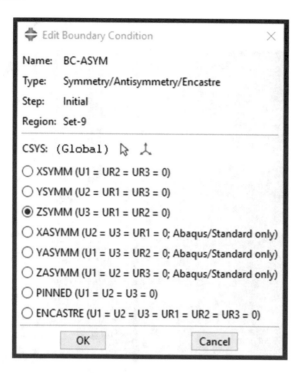

Fig. 6.80 Selecting the boundary condition

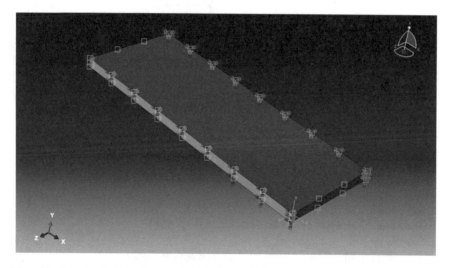

Fig. 6.81 Selecting the boundary condition in the workpiece

Fig. 6.82 Creating the
predefined field

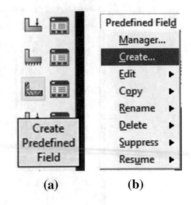

(a)   (b)

Fig. 6.83 Selecting the
temperature

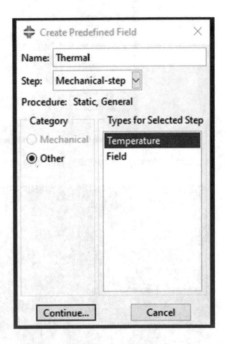

After that the job should be run.

Job → Manager (Fig. 6.95).

Then click on submit to run the simulation (Fig. 6.96).

Submit → Yes.

Visualization of the simulation results (Mechanical Model).

As can be seen, in the below figures the results for the thermal model can be seen
(Fig. 6.97).

To have a full view of the model, the asymmetrical part should be selected.

View → ODB display option → Select XY from the Mirror/Pattern bar → OK
(Figs. 6.98 and 6.99).

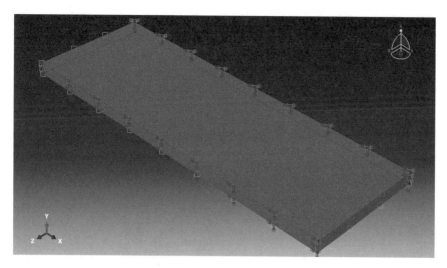

Fig. 6.84 Selecting the workpiece

Fig. 6.85 Confirming the set

Below Figures are the results for the FSW model (Figs. 6.100, 6.101 and 6.102).

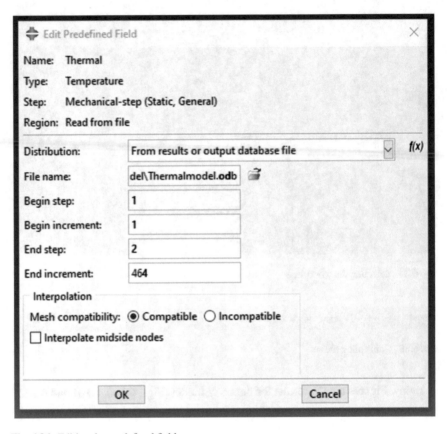

Fig. 6.86 Editing the predefined field

Fig. 6.87 Assigning the element type

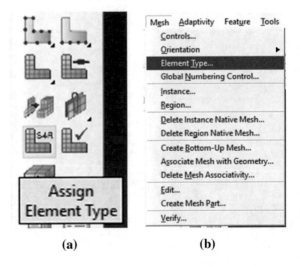

(a)   (b)

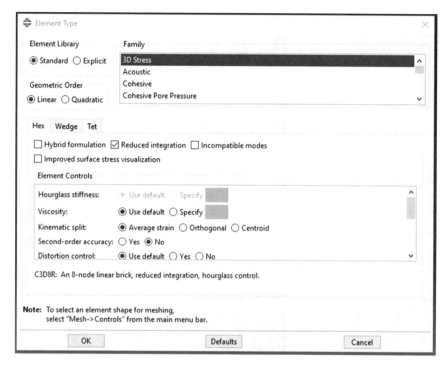

Fig. 6.88   Selecting the element type

Fig. 6.89   Selecting the field
output manager

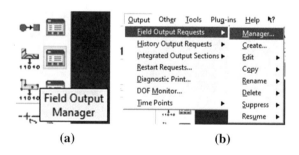

(a)   **(b)**

Fig. 6.90   Editing the field
output

Fig. 6.91 Editing the field output

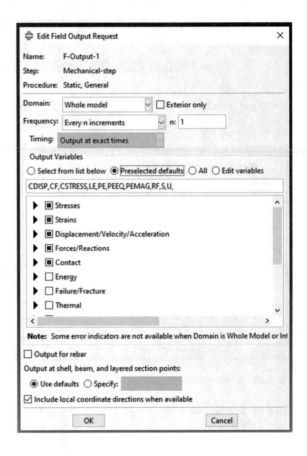

Fig. 6.92   Creating the job

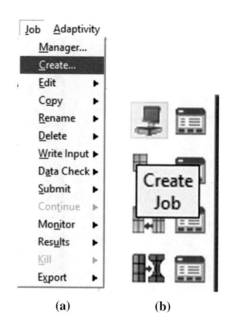

(a)   (b)

Fig. 6.93   Selecting the model

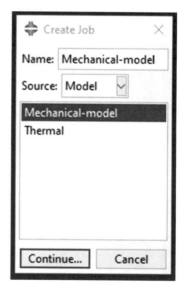

Fig. 6.94 Job creation

Fig. 6.95 Selecting the job manager

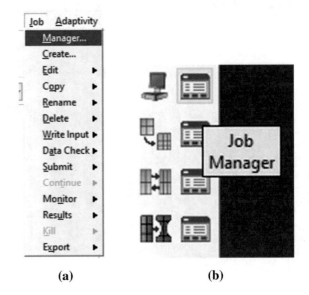

(a)   (b)

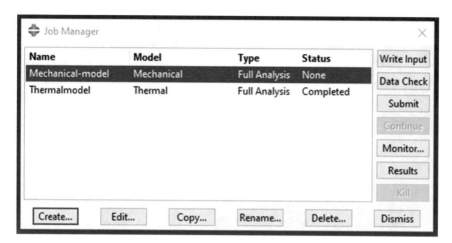

Fig. 6.96 Submiting the job

Fig. 6.97 Completed job

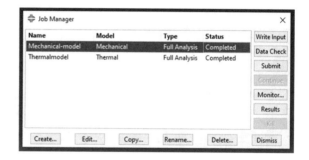

Fig. 6.98 Selecting the
ODB display option

View	Result	Plot	Animat
Save...			
Pan			F2
Rotate			F3
Zoom In/Out			F4
Box Zoom			F5
Auto-Fit			F6
Cycle Views			F7
Specify...			
Parallel			
Perspective			
✔ Show Model Tree			Ctrl+T
Full Screen			F11
Toolbars			▶
View Options...			
Graphics Options...			
Light Options...			
Image/Movie Options...			
ODB Display Options...			
Overlay Plot...			

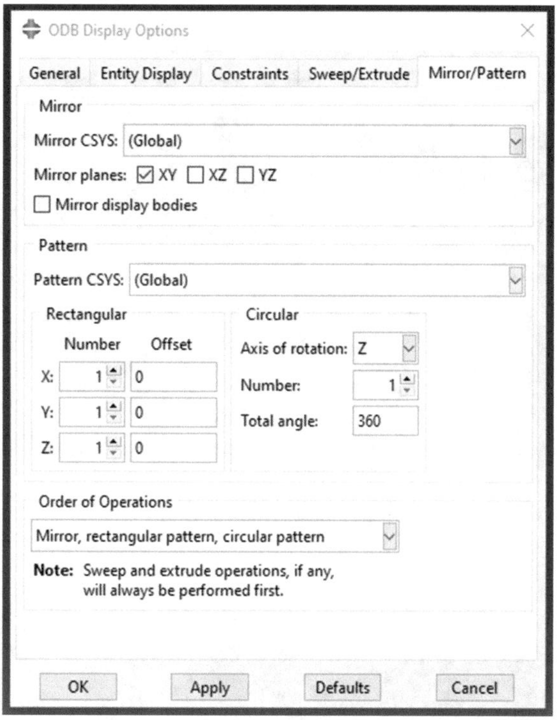

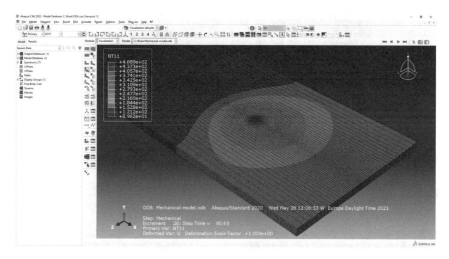

Fig. 6.99 Selecting the mirror tab

Fig. 6.100 Nodal temperature results

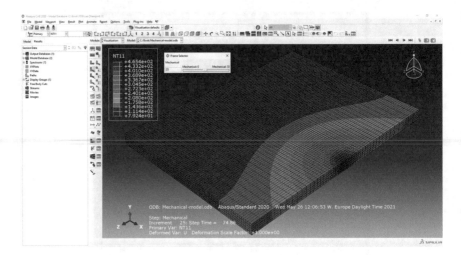

Fig. 6.101 Nodal temperature at the cross section

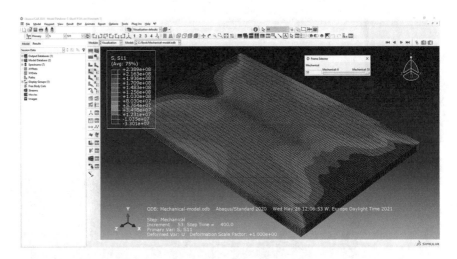

Fig. 6.102 Stress results

References

1. B. Meyghani, M. Awang, A comparison between the flat and the curved friction stir welding (FSW) thermomechanical behaviour. Arch. Comput. Methods Eng. **27**(2), 563–576 (2020)
2. B. Meyghani, M. Awang, C. Wu, Finite element modeling of friction stir welding (FSW) on a complex curved plate. J. Adv. Join. Process, **1**, 100007 (2020)
3. G. Casalino, *Recent Achievements in Rotary, Linear and Friction Stir Welding of Metals Alloys* (Multidisciplinary Digital Publishing Institute, 2020)
4. M. Awang, V. Mucino, Z. Feng, S. David, Thermo-mechanical modeling of friction stir spot welding (FSSW) process: use of an explicit adaptive meshing scheme. SAE Technical Paper 0148-7191 (2005)

Conclusions

Finite element methods (FEMs) is known as one of the most significant and useful computational techniques for solving various engineering problems. In this regard, various software in the field of mechanical, civil, mechanics, thermodynamics, electronics, etc. are employed. Among them, ABAQUS is known as one of the most famous and user friendly finite element analysis software. Friction stir welding (FSW) is a relatively new technique for joining similar and dissimilar materials. In this process, there are three sources for the heat including the friction, pressure and large plastic deformations. This book aims in training the detail modelling of FSW process in ABAQUS finite element package and it can be used as a reference for students and engineers in different engineering fields.

© The Editor(s) (if applicable) and The Author(s), under exclusive license   123
to Springer Nature Singapore Pte Ltd. 2022
B. Meyghani and M. Awang, *Welding Simulations Using ABAQUS*,
https://doi.org/10.1007/978-981-19-1320-4

Printed in the United States
by Baker & Taylor Publisher Services